David Levy's Guide to Observing and Discovering Comets

DAVID H. LEVY
Jarnac Observatory, Arizona

CAMBRIDGE
UNIVERSITY PRESS

PUBLISHED BY THE PRESS SYNDICATE OF THE UNIVERSITY OF CAMBRIDGE
The Pitt Building, Trumpington Street, Cambridge, United Kingdom

CAMBRIDGE UNIVERSITY PRESS
The Edinburgh Building, Cambridge CB2 2RU, UK
40 West 20th Street, New York, NY 10011-4211, USA
477 Williamstown Road, Port Melbourne, VIC 3207, Australia
Ruiz de Alarcón 13, 28014 Madrid, Spain
Dock House, The Waterfront, Cape Town 8001, South Africa

http://www.cambridge.org

First published 2003

Printed in the United Kingdom at the University Press, Cambridge

Typefaces Lexicon No. 2 10/14 pt and Lexicon No. 1   *System* LaTeX2_ε [TB]

A catalog record for this book is available from the British Library

Library of Congress Cataloging in Publication data

Levy, David H., 1948–
    David H. Levy's guide to observing and discovering comets / by David H. Levy.
        p.   cm.
    Includes bibliographical references and index.
    ISBN 0 521 52051 7 (pbk.)
    1. Comets – Popular works.   2. Comets – Observers' manuals.   I. Title: Guide to observing and
    discovering comets.   II. Title.
    QB721.4 .L46 2003
    523.6–dc21   2002031547

ISBN 0 521 82656 X hardback
ISBN 0 521 52051 7 paperback

The publisher has used its best endeavors to ensure that the URLs for external web sites referred to in
this book are correct and active at time of going to press. However, the publisher has no responsibility
for the web sites and can make no guarantee that a site will remain live or that the content is or will
remain appropriate.

David Levy's Guide to Observing and Discovering Comets

David Levy has held a lifelong passion for comets, and is one of the most successful comet discoverers in history. In this book he describes the observing techniques that have been developed over the years – from visual observations and searching, to photography, to electronic charge-coupled devices (CCDs). He combines a history of comet hunting with accounts of the latest techniques, showing how our understanding of comets has evolved over time. The book is suitable as a practical handbook for amateur astronomers, from those who are casually interested in comets and how to observe them, to those who want to begin and expand an observing program of their own. David Levy draws widely from his own experiences of a lifetime of observing comets, describing how enthusiastic amateurs can observe comets and try to make new discoveries themselves.

David Levy has discovered 21 comets, eight of them using his own backyard telescopes. In collaboration with Eugene and Carolyn Shoemaker, he discovered Shoemaker–Levy 9, the comet that collided with Jupiter in 1994, producing one of the most spectacular explosions ever witnessed in the solar system. He is a contributing editor for *Sky & Telescope* magazine, Science Editor for *Parade* magazine, and is the author or editor of twenty-nine books. He won an Emmy in 1998 as part of the writing team for the Discovery Channel documentary, *Three Minutes to Impact*. He is currently involved with the Shoemaker–Levy Double Cometograph comet search program, based at the Jarnac Observatory in Arizona.

To Wendee, my wife, I love you –
and to our grandchildren
Summer and Matthew
May your comets always be bright
and beautiful

Contents

Color plates will be found between pages 86 and 87.

Acknowledgments

When Simon Mitton of Cambridge University Press suggested that I write a guide to observing and discovering comets based on my personal experience, I was immediately interested in the idea. It is not that I needed to write a book on comets (this one is my sixth), but the possibility of writing something based on the passion I have had for so long was tremendously exciting. Thank you Simon, for this good suggestion. Thanks also to Wendee, my wife, who immediately encouraged me to follow Simon's suggestion, and who has been extremely helpful throughout this book's formative process. Wendee also prepared the index.

Dean Koenig and Scott Tucker of Starizona, amateur astronomer Tim Hunter, Daniel Green of the Central Bureau for Astronomical Telegrams, and Joan-ellen Rosenthal have been helpful with suggestions. Finally, Bill Buckmaster of KUAT television has been very supportive throughout.

Introduction: a personal odyssey with comets

Comets are like cats; they have tails, and they do precisely what they want.

David H. Levy, 1996.[1]

Time has not lessened the age-old allure of the comets. In some ways their mystery has only deepened with the years. At each return a comet brings with it the questions which were asked when it was here before, and as it rounds the Sun and backs away toward the long, slow night of its aphelion, it leaves behind with us those questions, still unanswered.

To hunt a speck of moving haze may seem a strange pursuit, but even though we fail the search is still rewarding, for in no better way can we come face to face, night after night, with such a wealth of riches as old Croesus never dreamed of.

Leslie C. Peltier, 1965.[2]

How many of us have looked up at the sky, and marveled at its supposed permanence? The sky on a clear night is one of the most reliable aspects of our lives; at a certain time on a particular date, we *know* that the stars will form their special patterns. This is almost true: Occasionally a comet will appear, interrupting this cosmic serenity and reminding us that even the heavens offer surprises, even the heavens are not immutable.

This book is designed to give its readers a sense of how people go about discovering and observing comets. The approach it follows combines a history of the field with the latest techniques for finding and observing comets. You can dip into the book at any part or chapter you like, but if you read the chapters as written, you will follow a natural progression of how our understanding of comets has evolved over time.

This is my sixth book about comets, a subject that continues to play a vital role in my own life.[3] That interest can be traced back to an airplane vapor trail

Figure I.1. On March 23/24, 1996, Canadian amateur astronomer Leo Enright took this photograph of Comet Hyakutake (C/1996 B2) at Sharbot Lake, Ontario, using a Canon 135 mm at f/2.5, with Kodak Ektachrome 1600 Professional film. Gamma Bootis is the brightest star inside the comet's tail. Courtesy Leo Enright.

on a sunny afternoon at the Hebrew school in Montreal. We kids were sitting around and chatting, waiting for our teacher to arrive, when I saw what looked like a comet with a bright trail hanging in the western sky. Since it did not appear to move, I thought it could not be an airplane. Could it be a comet, I recall thinking, and could it even be that I had discovered a comet? For a few minutes I was pretty excited, but then Mr. Mushkin arrived, our class began, and I forgot about the great Levy comet of 1959. I guess it was not a comet, for something bright enough to be seen in daylight would have surely made the annals of observing history. It might have been a plane moving directly toward or away from me. Frankly, to this day I am unsure what it really was.

Several months later, in elementary school, our sixth-grade teacher assigned each of us to give a 3-minute lecture in class. Terrified to talk in front of an audience, and especially in front of our strict teacher Mr. Powter, I spent many hours thinking not about a choice of topic but about how I could try to present a speech without actually having to face my sixth-grade peers. When speech day came, however, I was prepared. I presented a verbal tome on comets, but I used no notes. I had memorized everything I wanted to say, but in order not to have to face the audience, I carried with me a blank sheet of paper. Since my source was Leon Housman and Jack Coggins's *The Big Book of Stars*, a children's book I still have,[4] I can recall what I said that faroff day. Comets, I intoned, consist of a nucleus, a coma, and a tail – all still correct, although the 100+ mile diameter I quoted for a typical comet nucleus was off by a factor of ten. I mentioned Halley's comet, then on its way for a future rendezvous with Earth – its 1986 visit seemed a very long time in the future back then.

I also remember saying – possibly in reference to my own observation from a year earlier – that it was possible to discover comets, and that some people spend many years searching through small telescopes in the hope that they might, some day, find a new comet. Still holding my blank sheet of paper, I was tremendously relieved when the presentation was over. After the class applauded politely, Mr. Powter said, "Great speech, Levy. Can I see your notes?" The children in the front row, who easily saw my blank paper ploy, broke out in laughter.

That was March, 1960. By the end of that summer I was a stargazer committed to spending the next several years learning as much as I could about all aspects of astronomy. Comets were a part of those frenetic early years in astronomy, but my specific interest in them did not really return until October 1965, when I read about the discovery of a comet by two Japanese amateur astronomers, Kaoru Ikeya and Tsutomu Seki. The comet was a sungrazer, headed straight toward a rendezvous with the Sun during which it would complete a hairpin turn some 300000 kilometers from the Sun's photosphere.

The anticipation of seeing this wondrous comet really ignited my passion for comet hunting. That fall, while walking to an early French oral examination in tenth grade, and knowing that one of the questions would pertain to my choice of hobby, I decided to respond "Je veux découvrir une comete." It felt like the right thing to do at the time, and I especially enjoyed planning a search strategy. I knew that the best place for a visual, amateur search would be in the evening, after dark in the west, or in the morning, before dawn in the east; comets tend to be brighter when they are close to the Sun. However, since clouds often block the Montreal sky, my early searching took place whenever I could find a clear night. It was 19 years before I found my first new comet in 1984. Since then I have found, independently or with others, 20 other new comets.

Nothing in all those early years could have prepared me for what happened in the momentous spring of 1993. Observing with Gene and Carolyn Shoemaker, I took two photographs that recorded the motion of Comet Shoemaker–Levy 9 just a few months after an encounter with the tidal force of Jupiter that tore the comet into a string of fragments. This completely disrupted object looked like a squashed comet, as Carolyn described it, but through larger telescopes it resembled a string of pearls as its 21 pieces moved through the sky. On May 22, 1993, the International Astronomical Union announced that the comet, named Shoemaker–Levy 9, would collide with Jupiter in July 1994. Humanity learned from the catastrophic impact that followed, for as Comet Shoemaker–Levy 9 struck Jupiter, it provided an important lesson in our heritage: The impact recalled a solar system that had a violent youth, with comets plummeting into planets, bringing with them the elements of organic materials – carbon, hydrogen, oxygen, and nitrogen. On at least one of those planets, life arose from those chemicals.

A childhood whim and a grade-school assignment later developed into my lifelong passion for comets. In this book, I share the observing techniques that have been developed over the years by people who have enjoyed watching comets as much as I have. These techniques span the gamut from visual observations and searching, to photography, to electronic charge-coupled devices (CCDs). For visual observing, I decided years ago that if I planned to discover a comet, I should know what comets look like by observing as many known comets as possible. In 1985, I leaped from visual to CCD observing by joining the Near-Nucleus Studies Net of the International Halley Watch. As part of that effort, Steve Larson and I used a CCD to build a nightly record of the apparition of Halley's comet from late 1985 until 1989, and in the course of that survey recorded many other comets as well.

My work with conventional photography of comets began in 1970, when I photographed Comet Bennett. In 1988 I added photography to my search and observation program, and a few months later joined Gene and Carolyn Shoemaker's photographic search program. Over their careers together, the Shoemakers exposed more than 26 000 films. Being a part of that effort gave me a sense of what photography can accomplish while observing comets and searching for them. Now my wife Wendee and I, along with Carolyn Shoemaker, are part of a group called the Jarnac Comet Survey. We are searching in all three ways – visually, photographically, and with CCDs, and we try to record the known comets as they make their way through the sky. Just as I did in sixth grade, I still think that writing about a subject is a great way to learn about it. Through writing this book I hope that our group will be able to refine and improve its observations. At the same time, I hope that you, as a reader, will get an idea of what comet observation is all about. Whether you are just casually interested in comets and how you observe them, or if you want to begin or expand an observing program of your own, I hope that this book will give you a sense of the passion that we comet observers have.

NOTES

1. *The Today Show*, NBC Television, March 23, 1996.
2. Leslie C. Peltier, *Starlight Nights: The Adventures of a Star-Gazer* (New York: Harper & Row, 1965), 231.
3. The first five books are:

   Observe: Comets, with S. J. Edberg. (Astronomical League, 1985).
   An Observing Guide for Comets, Asteroids, Meteors, and Zodiacal Light, with Steve Edberg. Revised and expanded edition of *Observe: Comets* (Cambridge: Cambridge University Press, 1994).
   The Quest for Comets: An Explosive Trail of Beauty and Danger (New York: Plenum, 1994. Paperback editions from New York: Avon Books, 1995, and Oxford: Oxford University Press, 1995).
   Impact Jupiter: The Crash of Comet Shoemaker–Levy 9 (New York: Plenum, 1995).
   Comets: Creators and Destroyers (New York: Simon & Schuster, 1998).

   There is also a slide set:

   Comet Shoemaker–Levy 9 Slide Set, with the editors of *Sky & Telescope* (Cambridge, MA: Sky Publishing, 1994).

4. Leon A. Housman and Jack Coggins, *The Big Book of Stars* (New York: Grosset & Dunlap, 1955).

PART I

Why observe comets?

1

Of history, superstition, magic, and science

When beggars die there are no comets seen;
The heavens themselves blaze forth the death of princes.

Shakespeare, *Julius Caesar*[1]

When Shakespeare wrote that comets import change of times and states, he had something else in mind other than a comet literally plowing into Earth, with devastation so great as to destroy most of life here. No larger than a village, a comet moves lazily around the Sun, brightening and becoming more active as it closes in from a place beyond Jupiter, past the orbits of Mars, close to the orbit of Earth. Those of us who saw two spectacular comets in 1996 and 1997 will not soon forget those almost fearsome sights in the heavens. In March, 1996, the first of those two comets, Hyakutake, sported a filmy tail that stretched across the entire sky. The sight was remarkable, even in our time when we supposedly understand what a comet is and how it orbits the Sun. Past cultures, dating back to biblical times, were terrified by appearances so unusual that those who viewed them kept detailed records of their paths across the sky. "A comet appeared in the heavens like a twisting serpent," wrote Nicetus in 1182, "now writhing and coiling back upon itself; now it would terrify people with its gaping mouth; as if lusting for human blood, it seemed about to slake its thirst."[2] As late as 1528, Ambroise Pare wrote of a comet:

> So horrible was it, so terrible, so great a fright did it engender in the
> populace, that some died of fear; others fell sick … this comet was
> the color of blood; at its extremity we saw the shape of an arm holding
> a great sword as if about to strike us down. At the end of the blade
> there were three stars. On either side of the rays of this comet were
> seen great numbers of axes, knives, bloody swords, amongst which
> were a great number of hideous human faces, with beards and hair all
> awry …[2]

Now we keep records for different reasons. We want to learn about comets, and their orbits, and most especially we want to track those comets that could someday pose a threat to our planet.

A comet in the Bible?

Humanity's relation with comets dates back as far as historical records take us. This biblical passage from 1 Chronicles appears every year in the Passover Seder: "And David lifted up his eyes, and saw the angel of the Lord stand between the earth and the heaven, having a drawn sword in his hand stretched out over Jerusalem."[3] It describes some "sign" that protested an ill-advised census King David had ordered for his city of Jerusalem. Could that sign have been a comet? The ancient Hebrews, like their Arabic neighbors, enjoyed looking at the night sky and sought meaning among its many stars and events. A bright comet, appearing once every two decades or so, would have attracted their attention as much then as now.

Could King David have witnessed the comet that was apparently observed in Leo in 1002 BC? Or might it have been the comet that appeared half a century later, in the northern sky, around 959 BC. For no special reason other than the timing being about right, I like to think that the 959 BC comet was the comet of David.

Broom stars and bushy stars

It is a good thing that people throughout history have been moved by the passages of comets. Had they been less interested, we would not have such detailed records of their paths across the sky, as well as what they looked like and how their appearance changed with time. We have records of comets dating back to 1059 BC, when a comet with a tail pointing to the east appeared in the evening sky. Chinese recorders eventually noted two types of comet, the *po* and the *hui*. The *po*, or bushy star comet, had large fuzzy "coma" or atmosphere. If such comets had tails, they were unremarkable. The *hui* or broom star comet, on the other hand, was noted especially for its tail. Centuries later, observing from a different time and place, the Greek philosopher Aristotle divided comets into two classification groups: tailed and tailless. The tailless variety he called fringed and bearded stars. However, Aristotle did more than offer descriptions: He attempted to define the *nature* of comets. He thought that they formed when the Earth exhales hot, dry air into the upper reaches of its atmosphere. This view lasted for so long that it became almost impossible to challenge.

The heavens blaze forth the death of princes

Our understanding of comets as portents lasted a very long time. The ancient Romans feared them, and at least some in the audiences who came to watch Shakespeare's *Julius Caesar* 1500 years later still feared them. Shakespeare invoked comets and their supposed effects frequently in his plays. In *Julius Caesar*, Calpurnia begs her husband to stay away from the Senate. When Caesar asks why, she explains:

> Caesar, I never stood on ceremonies,
> Yet now they fright me. There is one within,
> Besides the things we have heard and seen,
> Recounts most horrid sights seen by the watch.
> A lioness hath whelped in the streets;
> And graves have yawn'd and yielded up their dead;
> Fierce fiery warriors fight upon the clouds,…
> And ghosts did shriek and squeal about the streets.
> O Caesar, these things are beyond all use,
> And I do fear them![4]

When all these events still did not move Caesar, Calpurnia added the appearance of comets in the night:

> When beggars die there are no comets seen;
> The heavens themselves blaze forth the death of princes.[1]

Did Calpurnia actually see a comet? Possibly she did in real life, but not *before* Caesar was murdered on the Ides of March, 44 BC at the foot of Pompey's statue. Two months later, during a series of games, a bright comet with a tail perhaps 12 degrees long – half the length of the Big Dipper (or Plough) – moved out of the northern sky. Plutarch wrote "among the divine portents there was also the great comet; it appeared very bright for seven nights after the murder of Caesar, then disappeared."[5] Calpurnius Siculus went further, blaming the comet for the civil war that followed: "when, on the murder of Caesar, a comet pronounced fatal war for the wretched people."[5]

To the people of Caesar's time, as to some in Shakespeare's time, comets were portents. Yet from this fear and attention came the beginnings of wisdom in the mind of a member of Emperor Nero's government. He was the writer Lucius Annaeus Seneca, and he lived in Rome in the first century AD. His writings, particularly his *Quaestiones Naturales*, made him immortal, but his life was ended at the whim of Emperor Nero. One chapter of the *Quaestiones*, called *De Cometis*, is a priceless look into the past of what people thought about comets.

Figure 1.1. A view of Comet Hyakutake (C/1996 B2) showing the comet behind a silhouette of a saguaro cactus plant, taken by the author on March 26, 1996, in Arizona, using a Yashika twin-lens camera.

Born around 4 BC, Seneca lived through the reigns of Caligula, Claudius, and Nero. When 17-year-old Nero became Emperor of Rome, Seneca, as his tutor, enjoyed considerable power in the government. In AD 59, Nero murdered his mother, and then sought Seneca's forgiveness. As Nero's tyranny set in, Seneca struggled to retain the Emperor's favor by using celestial events, like comets, to excuse his leader: A year after the murder, he invoked the comet of AD 60: "There is no reason to suppose," he wrote, "that the recent [comet] which appeared during the reign of Nero Caesar – which has redeemed comets from their bad character – was similar to the one that burst out after the death of the late Emperor Julius Caesar, about sunset on the day of the games to Venus Genetrix."[6]

It is amazing that Seneca found the time and energy to complete *De Cometis* in AD 60, while Nero was sinking into madness. "No man is so utterly dull and obtuse," Seneca wrote "with head so bent on earth, as never to lift himself up and rise with all his soul to the contemplation of the starry heavens, especially when some fresh wonder shows a beacon-light in the sky."[6]

Seneca had gifted insight about comets. "Blind to all the celestial bodies, each asks about the newcomer; one is not quite sure whether to admire or to fear it. Persons there are who seek to inspire terror by forecasting its grave import. But by my honour, no one could embark on a more exalted study."[6] Exalted, perhaps, but somewhat intolerant: Seneca viciously argued with those who, even though they had lived centuries earlier, had different opinions. "It requires no great effort to strip Ephorus [who lived more than three centuries earlier] of his authority; he is a mere chronicler."[6] Seneca accuses Ephorus of careless reporting: he says the Greek astronomer "asserts that the great comet [possibly the one of 373 BC] which, by its rising, sank Helice and Buris, which was carefully watched by the eyes of the whole world since it drew issues of great moment in its train, split up into two stars; but nobody besides him has recorded it."[6] Seneca could not imagine comets splitting into two or more pieces; but today we know that comets split with surprising frequency. Byzantine records show that around 822 "A comet was seen in the sky as a sort of two moons joined together brightly, and moreover separated by different attachments...."[7] Comet Biela divided into two pieces in 1846, and returned as a pair of comets a few years later, although it was never seen again. Comet West broke into four pieces in 1976. Comets Levy 1998e and Shoemaker–Holt, which were a single comet before they divided some 12000 years ago, were independently discovered by me and by the Shoemakers and Henry Holt in 1988. One of the many comets discovered by Project LINEAR, a modern automated comet and asteroid search, split apart in 2001. Lastly, the most famous and most recent example of a split comet was Shoemaker–Levy 9, which in 1993 divided into 21 fragments after an encounter with the tidal force of Jupiter, and was the next year destroyed by its collision with that planet.

As to the nature of the comets he loved, Seneca agreed with Aristotle that they are formed "by very dense air, and since the most sluggish air is in the north, they appear in greatest number in that direction."[8] Although comets are distributed across the sky almost at random, in Seneca's own experience, comets happened to favor the sky in the north. The comet of June AD 54, widely blamed for the death of Nero's predecessor Claudius, passed through

Figure 1.2. Comet Shoemaker–Levy 9 (D/1993 F2) as imaged through the NASA Hubble Space Telescope in January, 1994. This is one of the first images taken to test the telescope after it was repaired at the end of 1993. It shows the comet split into at least 21 fragments. NASA/HST.

the northern constellation of Gemini. Another comet the following year also appeared in the sky north of Cancer, and the one in AD 60, shortly after the beginning of Nero's reign, might have passed near the north celestial pole. Seneca's major contribution to the *science* of comets was that even though he saw them as atmospheric creations, he considered them permanent. "I rank it among Nature's permanent creations," he declared:

> "In none of the ordinary fires in the sky is the route curved; it is distinctive of a star [meaning a planet] that it describes a curve in its orbit. Whether other comets had this circular orbit I cannot say. The two in our own age [the comets of AD 54 and AD 60] at any rate had…. A comet has its own settled position. For that reason it is not expelled in haste, but steadily traverses its course; it is not snuffed out, but takes its departure."[9]

A major contribution is Seneca's invocation of the work of earlier thinkers; in fact his summary of the views of Apollonius of Myndos, the fourth-century BC scholar, might be the only record of those views. A comet, Apollonius had thought, is "a distinctive heavenly body, just as the sun or the moon is."[9] He even explained how comets brighten as they approach Earth, then fade as they depart. Four centuries before Seneca, Apollonius had hit upon the real nature of comets, only Seneca did not see it that way. If comets really brightened as they drew closer to Earth, then why, he asked, are some comets at their brightest when they first appear on the scene? Considering that Seneca did not have the benefit of our modern understanding of the nature of comets, his argument makes sense. Had Apollonius had access to a modern telescope, and to the mathematics of computing orbits, he would have countered that some comets might approach from behind the Sun, brightening as they arrive, but entirely unseen until they suddenly appear at their maximum brightness.

How wonderful it would be if we could gather all the great cometary thinkers, and have them debate comets! Imagine Seneca lecturing Apollonius, and then Apollonius getting the last word. Seneca's reality, sadly, was far darker. Years later the Roman historian Tacitus would look back on the days of Nero and his reaction to the comet of the summer of AD 60 – "a phenomenon which, according to the persuasion of the vulgar, portended change to kingdoms: hence, as if Nero had been already deposed, it became a topic of inquiry, who should be chosen to succeed him."[10] Seneca was quite obviously still trying to stay on Nero's good side, but as Tacitus pointed out, the appearance of the comet of AD 60 was punctuated by another sign that hit close to home: "as Nero sat at meat in a villa called Sublaqueum, upon the banks of the Simbruine lakes, the viands were struck by lightning and the table overthrown…."[10]

In the year AD 65, Nero accused Seneca of participating in a attempted coup and ordered him to "prepare for death" – according to custom this gave Seneca his choice of demise. Seneca chose cutting his wrist and bleeding to death, leaving *Quaestiones Naturales*, his priceless treasure, to be lost for more than a thousand years. Finally, in the twelfth century his book was discovered. Two millennia after he wrote it, Seneca's towering contribution still allows us to make sense of how thinking about comets has evolved over time.

NOTES

1. *Julius Caesar*, 2.2.30–31 (This and subsequent Shakespeare quotations are taken from *William Shakespeare: The Complete Works*, ed. Peter Alexander (London and Glasgow: Collins, 1964).
2. Lucien Rudaux and Georges de Vaucouleurs, *Larousse Encyclopedia of Astronomy* (New York: Prometheus Press, 1959), 241.
3. 1 Chronicles 21 : 16.
4. *Julius Caesar*, 2.2.13–19, 24–26.
5. A. A. Barrett, "Observations of comets in Greek and Roman sources before A.D. 410," *Journal of the Royal Astronomical Society of Canada*, 72:2 (1978), 81–106.
6. Seneca, Lucius Annaeus, *Quaestiones Naturales*, VII, *De Cometis* I, 1, translated by John Clark and Sir Archibald Geikie (London: Macmillan, 1910).
7. R. F. Rodgers, "Newly-discovered Byzantine records of comets," *Journal of the Royal Astronomical Society of Canada*, 56:5 (1952), 177–180.
8. Seneca, *De Cometis* XXIII, 1.
9. Seneca, *De Cometis*, XVII, 1.
10. *The Works of Tacitus*, Oxford translation (London: G. Bell & Sons, 1910), 367–368.

2

Comet science progresses

"Nova stella, novus rex!"

Bayeux Tapestry inscription above Halley's Comet

Ancient peoples did not know it, but a single remarkable comet, returning again and again, has punctuated human history for thousands of years. Although the English astronomer Edmond Halley had determined by the middle 1700s that the comets of 1531, 1607, and 1682 were actually separate visits by the same object, this comet has been playing on humanity's interests and fears for centuries. In a fortuitous series of coincidences, Halley's Comet (see Plate II) appeared at such critical moments as the defeat of Attila the Hun in 451 AD, and the Norman conquest of England in 1066. It was that latter visit that underscored the perception that the heavens themselves still blazed forth the death of princes. "*Nova stella, novus rex!*" ("New star, new king!") was the battle cry on the beautiful tapestry at Bayeux that later wove the story of that battle. By the fifteenth century that maxim still had not changed. A story from *The Illustrated London News* tells the story of a comet, although not Halley's, whose untimely visit caused the death of a prince. The guilty comet appeared in 1402 and was visible in broad daylight for a week:

> There is no doubt, however, that comets sometimes really did produce fatal effects. In June, 1402, one appeared in Italy which literally killed the famous John Galeas Visconti. The astrologers of the Prince had predicted that his death would be announced by a comet of extraordinary magnitude, and the celestial phenomenon had no sooner become visible than his Highness, speechless from fright, sank to the ground and died.[1]

A genius of the highest order

When the breakthrough finally came that allowed us to see comets as full-fledged members of the solar system, it had nothing to do with their appearance or structure. It had everything to do with their orbits, or paths around the Sun, and with the genius of one man, Edmond Halley. Born November 8, 1656, this son of a soapmaker and salter had a keen interest in science. As a 20-year-old student at Oxford he published his first paper. He was not one to stand on ceremony, however; impatient with Oxford, he dropped out and headed south that same year to the island of St. Helena, the island that, more than a century later, would serve as Napoleon's exile home after the Battle of Waterloo.

Halley's mission to that south Atlantic island came about because he realized that the southern sky was virtually unexplored. As a result of his visit, he charted many objects that were completely invisible from England. The observations he made resulted in *Catalogus Stellarum Australium*, a catalog of stars in the southern hemisphere that the young astronomer published in 1678. Halley's mentors at Oxford were so delighted with the catalog that they allowed him to reenter Oxford, and later awarded him a master's degree without requiring him to take the exams. Coupled with observations made later in his long life, Halley would discover that stars do change their positions relative to each other. He also inspired expeditions around the world to see Venus in two of its rare transits as the planet passed in front of the Sun in 1761 and 1769; the expeditions were arranged in response to his suggestion that observations of these transits could allow a precise calculation of the distance between Earth and the Sun. We will remember Halley again during the forthcoming transits of Venus that take place in 2004 and 2012.

With his increasing passion for the night sky, Halley turned his attention to the subject that would ensure his legacy and fame: Those objects that move through the sky at supposedly irregular intervals, the comets. Thinking that discovery of the true nature of comets lay in their orbits, Halley decided to study the routes taken by the comets he saw in 1680, 1682, and 1683. Using the accurate observations made by the astronomer John Flamsteed, Halley calculated the orbits of these comets. His successful completion of this task inspired him to work out the orbits of 21 other comets that had appeared from 1698 all the way back to 1337. As he pored over his results, he was surprised to learn that three of these comets had almost identical orbits. He quickly noted also that these comets appeared at intervals of roughly 76 years: In 1531, 1607, and 1682. An excited Halley shared these results with his friend Isaac Newton, who agreed with Halley's conclusion that these three comets were probably returns of the same comet. Halley was proud of his work: "Wherefore if according to what we have already said it should return again about the year 1758," he wrote immodestly, "candid posterity will not refuse to acknowledge that this was first discovered by an Englishman."[2] According to Brian Marsden, a fellow

Englishman and celestial mechanician at Harvard, who uses modern computers to calculate the orbits of hundreds of comets, Halley "really had the benefit only of the cometary orbits he computed, not so much the records of the comets themselves at roughly 76-year intervals."[3]

Although Halley suspected that the comets of 1378 and 1456 were earlier appearances of the same comet, he was not certain of this. There had also been another in 1301, but Halley thought this too early to be a return of this particular comet. It later turned out that the comet's orbital period varied from 74 to 79 years, depending on perturbations from the planets. It was not until the nineteenth century that celestial mechanicians like J. Russell Hind were able to connect earlier apparitions with Halley's comet, and early this century P. H. Cowell and A. C. D. Crommelin used ancient records to confirm the comet's visit as far back as 240 BC. It has since been shown that, over the centuries, the effect of "planetary perturbations" (now measurable) has been variable on different comets. For some, the effect would turn out to be negligible. For others, like Halley, the effects would be considerable and easily measurable. For Comet Shoemaker–Levy 9, as we shall see in the next chapter, the consequences would be catastrophic.

Back in the mid-1700s, Halley was uncertain just how severe the problem of planetary pulls on comets was. As he grew older, Halley couched his predictions in progressively less certain terms. In Halley's first prediction in 1705, notes Caltech's Donald Yeomans, Halley wrote in Latin that "I shall venture confidently to predict its return in 1758." When he translated that into English later that year, the prediction appeared as a far more modest "I dare venture to foretell…." Ten years later Halley hedged even more with "I think, I may venture to foretell … ;" and in his last writings he meekly said became "if … it should return again about the year 1758…."[4] Yeomans, one of today's most experienced orbit computers, believes that as Halley grew older he became more aware of just how complicated the calculation of comet orbits really is when the effects of the planets are taken into account. After all, he could predict all he wanted, but his comet was out there beyond the orbit of Neptune, and (anthropomorphically speaking) only that ball of ice and dirt could know precisely where it was and when it would return. Comets are like cats, I like to say: They both have tails, and they both do precisely what they want to do.

Halley died in 1742, after a full life of service to astronomy. He did not live long enough to see his forecast come true. During 1758, a veritable army of mathematicians and astronomers were frantically searching for the comet, on paper and in the sky. It was not until Christmas Night 1758 that Johann Georg Palitzsch, a Dresden farmer, spotted the comet with a small telescope; the French astronomer Charles Messier confirmed its appearance in January, 1759.

This achievement inspired the following generation to continue the mathematical search for other possibly returning comets. Six years later, in 1765, Nicolas-Louis de Lacaille first referred to the comet as Halley's Comet,[5] and for more than half a century, it stood alone as the only comet proven to be periodic.

Comet Halley's visit in 1759, followed so assiduously by the famous French observer Charles Messier, was a watershed for comet observing. When the comet returned in the year of Mark Twain's birth in 1835, astronomers and the public were more sophisticated about comets. Other comets had also been shown to be periodic. Anders Johan Lexell computed that a comet discovered by Charles Messier in 1770 had a period of only 5 years, but there was more: Lexell showed that this strange comet had come close to Jupiter just before its approach to Earth, and that afterward it approached Jupiter a second time. In this later encounter, Jupiter's gravity thrust the comet right out of the solar system. There is also the work of Johann Franz Encke, who linked the comets that appeared in 1786, 1795, 1805, and 1819, to fit the orbit of a single comet that would next return in 1822. Encke was right, and the comet that now bears his name has by far the shortest orbital period of any comet, a brief $3\frac{1}{3}$ years. I even made an independent discovery of Comet Encke while comet searching in 2000.

Comets and the origin of life

Halley's comet was sighted again from Heidelberg in 1909, the year my father was born. By the next year, when Mark Twain died, doomsayers and hustlers were distributing comet pills to ward off the supposed evil effects to be expected when Earth passed through the tail of Halley's Comet when it made a close pass in May of that year. In 1985, the year my father passed away, the European Space Agency was preparing its Giotto spacecraft to rendezvous with the comet's nucleus (see Plate II and Figure 2.1). The results from that March 1986 flyby were amazing. The spacecraft recorded that the abundances of the comet's "CHON" particles of carbon, hydrogen, oxygen, and nitrogen were almost identical to those in humans. Is this relation a cosmic coincidence? Most astronomers think not. During the early period following the formation of the solar system, Earth was pelted by comets that contained enough water to cover the surface of the planet to a depth of more than 20 feet! Comets might have contributed as much as ten times the present mass of water in all Earth's oceans. A good amount of this water would have been vaporized by the impacts themselves, but the supply seems to have been more than enough.[6]

Although comets were the major source of water in the primordial Earth, they were probably not the only source. Lava, which contains water, was carried

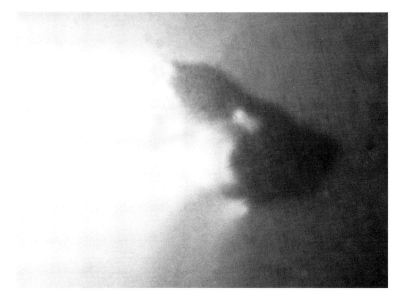

Figure 2.1. The nucleus of Halley's Comet (1P/Halley) as seen through the Halley Multicolour Camera, Giotto spacecraft, March, 1986. The figure is a composite of 60 images. Courtesy European Space Agency, Max-Planck-Institut für Aeronomie (H. U. Keller).

by volcanic eruptions from beneath the surface of Earth in quantities far greater than at present. Over the centuries, Halley's Comet has taught us two lessons about the nature of our solar system. First, comets travel around the Sun in orbits that bring them from the far reaches of space to the vicinity of Earth. By inference, if comets can get close to Earth, then they can occasionally collide with it. The second lesson is that comets probably left their precious organic materials on Earth after they collided, a process that led inexorably to the origin of life on this world.

NOTES

1. *Illustrated London News*, August, 1861.
2. Edmond Halley, *Astronomical Tables* (1749, [revised] 1752); cited in Donald K. Yeomans, *Comets: A Chronological History of Observation, Science, Myth, and Folklore* (New York: John Wiley, 1991), 122.
3. B. G. Marsden, personal communication, October 1, 1992.
4. Yeomans, 122–123.
5. Nicolas-Louis de Lacaille, *Mémoires de mathématique et de physique, tirés des registres de l'Académie Royale des Sciences, de l'année 1759* (1765) 522–544, cited in Yeomans, 138.)
6. Armand H. Delsemme, "Nature and history of the organic compounds in comets: An astrophysical view," in *Comets in the Post-Halley Era, Vol. 1*, eds. R. L. Newburn, Jr., M. Neugebauer and J. Rahe (Dordrecht: Kluwer Academic, 1991), 416.

PART II

Discovering comets

3

Comet searching begins

Seeing is in some respects an art which must be learnt. To make a person see with such a power is nearly the same as if I had been asked to make him play one of Handel's fugues upon the organ. Many a night have I been practising to see, and it would be strange if one did not acquire a certain dexterity by such constant practice.

William Herschel[1]

As it always has been throughout its history, Paris was one of the world's busiest places in the middle of the eighteenth century. *Ce va sans dire*, it was a very different place from today's Paris, but the Hôtel de Cluny still stands. In 1758 the astronomer Nicholas Delisle used it as observatory, along with his youthful assistant Charles Messier. At Cluny, Messier received astronomical training, which, combined with a good dose of passion, set the stage for history's first organized program for hunting comets.

Although Charles Messier was certainly not the first person to find comets, he was apparently the first to find them as a result of an organized search program. Others did precede him: Paolo Toscanelli, the late fifteenth century cartographer whose map of the Atlantic Ocean encouraged Columbus to set out on his voyages, also made careful drawings of six bright comets that appeared between 1433 and 1472. A man of eclectic interests besides cartography and astronomy, Toscanelli might have observed the sky so often that he saw these comets before others told him about them, but maybe he did not. Centuries later, it does not matter. What is important is that his drawings allowed later astronomers to compute the orbits of the comets, including that of the visit of Halley's Comet in 1456 that frightened Pope Calixtus.

Nor was Messier the first to discover a comet with a telescope; that honor seems to belong to Gottfried Kirch, who found a comet while bringing the tube of his telescope to point to Mars in 1680.[2] The comet he found later became bright enough to be observed with the unaided eye, but it was not a naked-eye object when he discovered it through his telescope.

The comet ferret

I have a hunch that Messier's comet search got its start because he failed to be the first person to see Halley's comet in 1758. Although Messier thought he was the first, that honor actually went to Johann Georg Palitzsch, who found the comet through his telescope on Christmas Night, 1758. More than 3 weeks later, Messier found the comet from Hôtel de Cluny on January 21, 1759. He was thrilled – "It was one of the most important astronomical discoveries," he wrote, "for it showed that comets could return."[3] He summoned Delisle, who observed the comet and then, for some unexplained reason, promptly ordered Messier not to announce it in any way. Delisle was not persuaded to announce Messier's recovery of Halley's Comet until April 1, a full 3 weeks after the comet had already rounded the Sun. (Locating a comet in this manner is called a "recovery" rather than a discovery.) By this time, Messier had already heard of Palitzsch's Christmas Night sighting and knew that he would have lost the race anyway, despite anything Delisle did.

Thus began Messier's lifelong love affair with comets. The very next year, 1760, he made his first of eight finds without competition from anyone else. He found comets in 1763, 1764, 1766, 1769, 1770, 1771, and 1773, before his chief rival Jacques Montaigne, a druggist in Limoges, France, snared his first in 1774 – the comet that in 1826 became known as Biela's Periodic Comet. This comet apparently split in two in 1846, returned as two comets in 1852, and then was never seen again; in 1872, when the comets should have appeared once more, there was instead a great storm of slow-moving meteors.

By 1781 Pierre Méchain had also joined the contest, but no other comet hunter outperformed Messier during his lifetime; in 1801, when he finally retired from comet hunting, Messier had discovered 12 comets.

Jean-Louis Pons's approach to searching

Although Messier discovered his last comet on July 12, 1801, he was not the first to spot it. A forty-year-old janitor found the comet near the Big Dipper (or Plough) a day earlier. This single comet was the first of 26 that now bear the name of Jean-Louis Pons.

Although Pons's name is borne by 26 comets, he actually found at least 30, and possibly as many as 37. However, to give Pons credit for 37 comets, we must count some discoveries in which he was not the first finder, several others for

Figure 3.1. Comet Levy 1990c (C/1990 K1) as imaged through the NASA/Hubble Space Telescope shortly after its launch in 1990. NASA/HST.

which he was far too late to have his name attached, two discoveries made years apart of the periodic comet now named after Encke, and comets that were never confirmed by other observers. Born in 1761, Pons was already an adult when he caught the bug of astronomy in 1789. In that year of the storming of the Bastille, the Marseilles Observatory hired him as a doorkeeper to keep watch on the observatory. Apparently he interpreted his job as keeping watch not only on the grounds of the observatory but also, with his small telescope, as standing guard over the sky with "le grand chercheur" – his favorite searching telescope. With the large 3-degree field of view of this telescope, Pons could cover large areas of the sky. He found a second comet in 1802, one in 1804 and another in 1806. While hunting with this telescope just before dawn on February 9, 1808, Pons saw a comet very close to the globular cluster Messier 12. Unfortunately he kept a rather poor record of this observation: in a field crowded with stars, the only three objects he drew were the new comet, Messier 12, and the nearby globular cluster Messier 10. Since no one else was able to observe this object, no orbit was calculated for it, and for over 180 years the comet remained unconfirmed. It turned out that a comet returning in 1902, Grigg–Skjellerup, was actually a return of Pons's comet. The 26th known periodic comet, it returns every 5 years. However, it was not until 1986 that the astronomer Lubor Kresák determined that 26P/Grigg–Skjellerup was indeed the comet that Pons had glimpsed fleetingly almost two centuries earlier.[4]

Despite his growing record of comet finds, for some reason Pons did not succeed in gaining the respect of his peers during the early part of his career. His rural upbringing, it is said, might have made him appear naive. Other astronomers teased him: After some time had passed since his most recent comet discovery, Pons asked the German astronomer Baron von Zach, then visiting from the Seeberg Observatory, for a hint on how he could find more comets. The astronomer thought for a few seconds and then suggested, preposterously, that Pons search more assiduously when there are lots of sunspots on the Sun. Half expecting that Pons would make a fool of himself and redesign his program to search the night sky when the sun sported large spots, von Zach was amazed to get a letter from Pons with profuse thanks. He had indeed observed large spots forming on the Sun, and soon afterward he dutifully found a new comet.[5] Although no correlation between comets and sunspots has ever been found, it actually makes some intuitive sense that when the Sun is more energetic, comets, which do react to solar energy, would brighten.

In 1811 Pons codiscovered a comet destined to be one of the best and brightest in history. Finding it 3 weeks after its discovery by the French observer Honoré Flaugergues, Pons can claim credit to its discovery in a time when communications were not nearly as good as now. The comet brightened quickly until it could be seen without a telescope, and it remained visible to the naked eye for 10 months. Its appearance was even credited with the coincidentally ultrafine wines from that year. Five years later, in October, 1816, the English poet John Keats compared the thrill he had experienced when reading Chapman's translation of Homer with that of discovering a new world:

> Much have I travel'd in the realms of gold,
>> And many goodly states and kingdoms seen;
>> Round many western islands have I been
> Which bards in fealty to Apollo hold.
> Oft of one wide expanse had I been told
>> That deep-brow'd Homer ruled as his demesne;
>> Yet did I never breathe its pure serene
> Till I heard Chapman speak out loud and bold:
> Then felt I like some watcher of the skies
>> When a new planet swims into his ken;
> Or like stout Cortez when with eagle eyes
>> He star'd at the Pacific – and all his men
> Look'd at each other with a wild surmise –
>> Silent, upon a peak in Darien.[6]

Most critics have written that Keats was recalling William Herschel's discovery of the planet Uranus in 1781, an event that occurred 14 years before he was born. I suspect instead that Keats had in mind the discovery, by Pons and others, of the comet of 1811.

With solid success behind him as a comet hunter, in 1813 Pons at last was promoted from doorkeeper to assistant astronomer, and in 1817 he became director of an Italian observatory close to the town of Lucca, near Florence. Pons later became director of Florence's Observatory. He found his last comet in August of 1827, and he died 4 years later. Only one person, Carolyn Shoemaker, has beaten Pons's record of 26 comets named for an individual; Shoemaker has 32.

The Herschels' approach to searching for comets

Having produced three astronomers who have left their mark in the sands of time, the Herschel family is one of the most famous in the history of astronomy. Born in Hanover, Germany, in 1738, Wilhelm Herschel was the son of a musician, and he became a composer and conductor as well as a skilled performer on the horn, harp, and organ. He used the English version of his first name, William, after he moved to England in 1757.

In May of 1773, Herschel made a small purchase that would have major consequences. He records in the May 10 entry of his diary: "Bought a book of astronomy and one of astronomical tables."[7] Soon his developing interest led to his construction of a 4-foot long refractor, but eventually he realized that reflectors offered a more useful design for him. He ground his own mirrors, made of a "speculum metal" complex "of 21 copper, 13 tin, and one of Regulus of Antimony, and I found it very good, sound white metal."[8] So enthusiastically did Herschel immerse himself in his zeal for telescopes that he moved to a larger home in Bath, England. His younger sister Caroline Lucretia, born in 1750, moved into it as well. She joined her brother in his work, and helped him set up the house's backyard as an observing site.

By 1781, only 8 years after he fell in love with the stars, Herschel was deeply involved in a systematic survey of the sky, studying and recording each star. "On Tuesday the 13th March," he wrote, "between ten and eleven in the evening, while I was examining the small stars in the neighborhood of H Geminorum, I perceived one that appeared visibly larger than the rest; being struck with its uncommon magnitude I compared it to H Geminorum and the small star in the quartile between Auriga and Gemini, and finding it so much larger than either of them, suspected it to be a comet."[9] Although the object was relatively

Figure 3.2. Comet Levy 1990c (C/1990 K1) as imaged through the 18-inch Schmidt camera at Palomar Observatory in California, by the author, in August, 1990.

bright, perceiving it not as a point of light but a disk was a real challenge even to a trained eye.

Herschel thought he had discovered a comet, but it was so peculiar that news of it spread rapidly to Messier, who observed the new object at every opportunity. "I am constantly astonished at this comet," he wrote to Herschel in the late spring of 1781, "which has none of the distinctive characters of comets, as it does not resemble any one of those I have observed, whose number is eighteen.... I have since learnt by a letter from London that it is to you, Sir, that we owe this discovery. It does you the more honour, as nothing could be more difficult than to recognize it, and I cannot conceive how you were able to return several times to this star – or comet – as it was absolutely necessary to observe it several days in succession to perceive that it had motion ... For the rest this discovery does you much honour; allow me to compliment you for it. I should be very curious, Sir, to learn the details of this discovery, and you will oblige me if you will be so good as to inform me of them."[10]

The answer to Herschel's riddle came on August 31, 1781, when the celestial mechanician Anders Lexell published a definitive orbit of Herschel's object. He declared that its path was almost circular, and that it never got closer to the Sun than 16 times the Earth's distance to the Sun. Herschel's object did not look much like a comet because it was not a comet. It was a planet, the first to be discovered in historic times.

The next year Herschel was appointed private astronomer to King George III, who insisted that he have the best telescope possible. Their first observing session, on July 2 at Windsor Castle, was a tremendous success. "My Instrument gave a general satisfaction," Herschel wrote to his sister; "the King has very good eyes and enjoys Observations with the Telescopes exceedingly."[11]

William Herschel's self-discipline and enthusiasm was infectious, especially to his sister Caroline. Shortly after her brother's great discovery, she began her own comet search program; her brother built her a 6-inch reflector for this work. In 1783 she discovered, low in the southern sky, the beautiful spiral galaxy in Sculptor now known as NGC 253. (I suggest, incidently, that this object be named Herschel's galaxy in her honor.) On August 1, 1786, when William was away in Germany, Caroline discovered her first comet. On the winter solstice of 1788, she discovered her second comet in Lyra. William was with her this time, and he described it as "a considerably bright nebula, of an irregular form, very gradually brighter in the middle, and about five or six arcminutes in diameter."[12] Although the comet's orbit was considered to be parabolic, meaning that the comet would not return, a century and a half later, in 1939, Roger Rigollet, of Lagny, France, found a comet that turned out to be Herschel's comet on its first return since discovery. Thus, this comet is now called Herschel–Rigollet, and it has a period of about 150 years.

The year 1790 was a banner year for Caroline. She found her third comet on January 7, a difficult find that only 3 days later became too close to the Sun to be seen. On April 18 she discovered her fourth, this time in the predawn sky in the constellation of Andromeda. She sighted a fifth comet on December 15, 1791; a few days before she discovered it, this comet had made a relatively close pass by Earth.

Caroline Herschel and Comet Encke

Another Caroline Herschel comet, found on October 20, 1805, turned out to be a rediscovery of a comet that Pierre Méchain had found on January 17, 1786. No one was aware of the connection at the time; nor were they aware of it 19 years later, when, on October 20, 1805, Pons discovered it for a third time. Meanwhile, when a German mathematician named Johann Franz Encke calculated an orbit for this comet, he was surprised to suggest that it might be a comet that returns every 12 years. On November 26, 1818, Pons discovered this same comet for a second time! Encke then connected Pons's new comet to the

1805 comet, but in so doing he realized that its period was not 12 years, but $3\frac{1}{3}$. Encke also calculated that Pons's comet was the same as those of Méchain and Caroline Herschel, confirming the orbit of this singular comet: no other known comet orbits the Sun as quickly as does Comet Encke.

Although Encke always believed the comet should be named for Pons, who discovered it twice, it is now named Comet Encke in honor of the mathematician who figured out its strange orbit. On June 2, 1822, the Australian searcher Charles Rumker recovered this comet, confirming Encke's work once and for all. Since then it has been followed on most of its returns. As we shall see in the next two chapters, it would be caught by Horace Tuttle in 1875, and would help Fred Whipple work out his "dirty snowball" theory in the 1950s. At its poorly placed return in 2000, I found it during the morning of August 9, about 100 times fainter than those earlier discoveries.

Caroline Herschel's seventh and final find, Bouvard–Herschel–Lee, was a bright, naked-eye object at its discovery in 1797, and fairly close to Earth. To honor this extraordinary achievement, the Royal Astronomical Society bestowed on her its highest honor, its Gold Medal, in 1828. By this time, Herschel was known internationally for her "eccentric vocation" that included a log of her discoveries fondly entitled *Bills and Receipts of my Comets*. During all this time, Caroline also continued to assist her brother; in fact she always placed a higher priority on his work than on her own search for comets. During one of her nightly roles as night assistant, Caroline suffered a broken ankle. When Caroline apparently ignored William's request to move the telescope in a particular direction, the story goes, William repeated, "Lina, move the telescope!" "I'm hooked!" Caroline yelled back. A large hook used for pulling the telescope along had struck her ankle. Such are the hazards faced by those passionate people who devote their careers to studying the night sky.

NOTES

1. W. G. Hoyt, *Planets X and Pluto* (Tucson: University of Arizona Press, 1980), 12.
2. Gary W. Kronk, *Comets: A Descriptive Catalog* (Hillside, NJ: Enslow Publishers, 1984).
3. K. G. Jones, *Messier's Nebulae and Star Clusters*, Second Edition (Cambridge: Cambridge University Press, 1991), 347.
4. Daniel Green, *International Comet Quarterly*, 8, 111 (1986).
5. R. K. Marshall, "Astronomical Anecdotes," *Sky & Telescope*, 3 (April, 1944), 19.
6. John Keats, "On First Looking into Chapman's Homer" in *Keats: Poetical Works* ed. H. W. Garrod (London: Oxford University Press, Second Edition, 1966), 38.
7. C. A. Lubbock, ed., *The Herschel Chronicle: The Life-Story of William Herschel and His Sister Caroline Herschel* (Cambridge: Cambridge University Press, 1933), 60.
8. Lubbock, 66.
9. Herschel announced his discovery at the end of March, 1781, to the Bath Literary and Philosophical Society. This "Account of a Comet" appears in *The Scientific Papers of*

Sir William Herschel, Vol. I, ed. J. L. E. Dreyer (London: The Royal Society and the Royal Astronomical Society, 1912), 30–38.

10. Charles Messier to William Herschel, in Lubbock, 86.

11. W. Herschel to C. Herschel, July 3, 1782, in Lubbock. See also "America's last King and his observatory" in J. Ashbrook, *The Astronomical Scrapbook: Skywatchers, Pioneers, and Seekers in Astronomy* (Cambridge, MA: Sky Publishing Corporation, 1984), 17.

12. Kronk, 264.

4

Tails and trails

– I am like a slip of comet,
Scarce worth discovery, in some corner seen
Bridging the slender difference of two stars,
Come out of space, or suddenly engender'd
By heady elements, for no man knows;
But when she sights the sun she grows and sizes
And spins her skirts out, while her central star
Shakes its cocooning mists; and so she comes
To fields of light; millions of travelling rays
Pierce her; she hangs upon the flame-cased sun,
And sucks the light as full as Gideon's fleece:
But then her tether calls her; she falls off,
And as she dwindles shreds her smock of gold
Amidst the sistering planets, till she comes
To single Saturn, last and solitary;
And then she goes out into the cavernous dark.
So I go out: my little sweet is done:
I have drawn heat from this contagious sun:
To not ungentle death now forth I run.

Gerard Manley Hopkins, 1864[1]

The nineteenth century saw the appearance of several major comets besides the one in 1811 that I suspect inspired Keats. When, in Windsor, New South Wales, on May 13, 1861, amateur astronomer John Tebbutt discovered a comet, he had no idea that within 6 weeks it would brighten to become one of the century's greatest; but he knew soon afterward, for unlike many of today's comet hunters, Tebbutt made a point of carefully observing the comets he found, even attempting the difficult task of calculating their orbits. At the end of June, the huge comet's head was in the northern sky near the bright star

Capella, while its tail stretched more than half way across the heavens to the constellation of Hercules. Since he kept journals for all his observations, we are able to read about Tebbutt's own remarkable observation of the tail of his comet: "In the evening of June 30 I observed a peculiar whitish light throughout the sky, but more particularly along the eastern horizon. This could not have proceeded from the moon, but was probably caused by the diffused light of the comet's tail, which we are very near to just now."[2] Earth, it turned out, went right through the outer reaches of the comet's tail on the night of June 30.

Energized by his major discovery and his comet's subsequent behavior, by 1863 Tebbutt had completed building his own small observatory by himself; it was a structure that would last for many years and include a transit instrument for accurately measuring comet positions. On the evening of May 22, Tebbutt spotted another comet, this one with the naked eye, in the southern constellation of Columba, the dove. "Immediately on its discovery," he recorded, "I obtained, with the $4\frac{1}{2}$-in. equatorial, eight good measures of the nucleus from one of the bright stars just mentioned. On the following day I notified the discovery to the Government Observatories of Sydney and Melbourne."[3] The Melbourne Observatory, incidently, was opened around that time to help out with the southern hemisphere section of an international multimillion star cataloging effort called *Carte de Ciel*, and also to observe the transits of Venus that took place in 1874 and 1882. When I visited the observatory on a beautiful clear day in November 2001, I was quite moved by this testament to the astronomical tradition that Australia has enjoyed for so long. I also thought of the excitement the observatory astronomers must have had in 1882 when they observed the transit of Venus, and also the rapidly brightening Tebbutt's comet. By the end of June the comet was of the first magnitude.

Communications were slow during Tebbutt's time, especially with regard to the comet of 1861, whose discovery went virtually unknown in England until, in late June of that year, it suddenly appeared over the southern horizon with a first-magnitude central part and a tail stretching over most of the sky. William Ellis, the observer assigned to England's Greenwich Observatory that night, saw the comet rise over the southern horizon. Anxious to observe it, he also feared that his employer, George Airy, the British Astronomer Royal, would fire him for not maintaining his prescribed sequence of observations. Torn between the work he had been assigned to do and the work he wanted to do, Ellis turned his telescope to the comet in secret.[4]

Of tails and trails

The nineteenth century was notable not just for what comets looked like, but also for the small but notable extras they brought with them. On the night of November 17, 1833, Earth passed through the dust trail of a small comet that had passed by undetected. For a few hours meteors fell at the rate of hundreds of thousands per hour. At the comet's following return in 1866, two amateur astronomers, Ernst Tempel and Horace Tuttle, searching independently, discovered it. Soon afterwards, Giovanni Schiaparelli, an astronomer later famous for his theory of canals on Mars, suggested that meteors are related to specific comets, and that the meteor storms of 1833 and 1866 came from this comet – Comet Tempel–Tuttle.

Like amateur astronomers throughout time, even those observatory employees who were not allowed to view the comet of 1861, nineteenth-century comet discoverers had to sandwich their recreational observing time between the demands of work, family, and country. Even as Tuttle discovered his fourth comet in 1863, Abraham Lincoln delivered his proclamation of emancipation, one of the signature events of the Civil War. Tuttle decided to join the Union Navy as acting paymaster.[5] In 1864 his ship, the *Catskill*, was anchored off Charleston, South Carolina, to blockade that city and Wilmington against confederate forces. After the war ended, Tuttle remained with the Navy. Three years after his discovery of the Leonid meteor comet, Tuttle was sent across the country to serve as paymaster aboard the monitor ship *Guard*. However, when he reported to his captain, he had misplaced the $8800.90 in pay for the ship's crew that he had been sent with. Believing that Tuttle had squandered these funds, the Navy charged him with theft and ordered him to appear in Washington for a court-martial. At the time in the midst of an embezzlement scandal, the Navy had just sentenced an army paymaster to life imprisonment in a federal penitentiary, a harsh punishment that President Grant later reduced to 5 years. Tuttle must have known about this highly publicized event, and probably feared that the same punishment might happen to him. That fear did not keep him from obtaining permission to use the Navy's 26-inch refractor just 3 days into his own trial. On January 23, 1875, he was the first observer to view the periodic Comet Encke as it returned that year. Three weeks later Tuttle was found guilty of embezzlement, and received the surprisingly light sentence of dishonorable discharge from the Navy. The action was approved by President Grant.[6] Tuttle never admitted to doing anything wrong, but what happened to the almost 9000 dollars, an extraordinary amount of money at the time, remains a mystery.

Edward Emerson Barnard and the hoax of the comet-seeker telescope

By the 1880s a new generation of observers, led by William R. Brooks and Edward Emerson Barnard, had taken over the art of comet hunting. These men found their first comets within a month of each other and were keen competitors; their career totals were 22 and 16 named comets, respectively. Born in Scotland, William Brooks moved with his family to New York in 1857, settling upstate in Phelps, where he became a photographer. He searched for comets with his home-made $9\frac{1}{4}$-inch reflecting telescope.

In the same year that Brooks moved to America, Barnard was born in Nashville, Tennessee. When he was less than 10 years old he left school (his formal education lasted for only 2 months) and became a photographer's assistant. Less than a month before Brooks found his first comet, Barnard found his, on September 17, 1881. There is no question that a $200 prize offered at the time for any comet discoveries propelled the competition during this period, especially for a man like Barnard, with a new wife and little wealth. He admitted:

> I had been searching for comets for upward of a year with no success, when a prize of two hundred dollars for the discovery of each new comet was offered by the founder of the Warner Observatory through the agency of Dr. Lewis Swift, its Director. Soon after this it happened that I found a new comet, and was awarded the prize. Then came the question, "What shall I do with the money?" After due deliberation it was decided that we would try to get a home of our own therewith. I had always longed for such a home, where one could plant trees and watch them grow up and call them our own. So we bought a lot with part of the money, which was on rising ground selected in part because it gave me a clear horizon with my telescope.
>
> After some saving and mainly a mortgage on the lot, we built a little frame cottage where my mother, my wife and I went to live. Those were happy days, though the struggle for a livelihood was a hard one, working from early to late, and sitting up the rest of the twenty-four hours hunting for comets. We looked forward with dread to the meeting of the bills which must come due. However, when this happened, a faint comet was discovered, and the money went to meet the payments. The faithful comet, like the goose that laid the golden egg, conveniently timed its appearance to coincide with the advent of those dreaded notes. And thus it finally came about the house was built entirely of comets. This fact goes to prove the great error of those scientific men who figure out that a comet is but a flimsy affair after all, infinitely more rare than the breath of the morning air, for here was a strong compact house, albeit a small one, built entirely of them. True, it took several good-sized comets to do it, but it was done, nevertheless.[7]

Quite besides the monetary award, comet hunting during the 1880s was characterized by one of the most impressive series of bright comets ever to parade through the inner part of the solar system. Compared especially with the picayune harvest we have had in recent years, the procession was truly superior. It began in 1880 when a large comet dominated the southern-sky constellations of Tucana, Grus, and Phoenix; its tail grew to 40 degrees in length to cover a quarter of the visible sky. On May 22 of the following year, as we have seen, John Tebbutt discovered a comet that would brighten to magnitude -14 (much brighter than the full Moon) as it reached perihelion in a hairpin dance around the Sun. David Gill, director of South Africa's Cape Observatory, described the rising of this comet:

> There was not a cloud in the sky, only a merging into a rich yellow that fringed the brackish blue of the distant mountains, and over the mountains and amongst the yellow an ill-defined mass of golden glory rose with a beauty I cannot describe.
>
> The Sun rose a few minutes afterwards, but to my intense surprise the comet seemed in no way dimmed in brightness, but becoming instead whiter and sharper in form as it rose above the mists of the horizon.[8]

Early in the morning of October 14, Barnard dreamt about a sky filled with comets, a dream so real and vivid that when he awoke it took a few minutes before he realized it was only a dream. He walked outdoors to see a dark sky and the comet just rising in the east. He studied the comet through his telescope before beginning his regular morning search for new comets. Moving his telescope to the southwest, he had not advanced more than 5 or 6 degrees before he found a group of a half-dozen small comets. Barnard was resuming his dream, but this time his dream was reality. The observation was confirmed the following night by observers in Europe; but although the little comets were traveling at the same rate and direction as the main body, they all disappeared in less than a day.

A week after Barnard's find, on October 21, Brooks found a companion comet several degrees northeast of the comet. Brooks's object also vanished. These "temporary comets" cannot be explained by the comet having broken into several pieces as it reached perihelion on September 17 – the comet's most vulnerable time as it interacts with the strong tidal forces of the Sun; nor could they have split off the main comet and moved so far from it in such a short time. Brian Marsden of Harvard suggests that they might have split off from the main comet at its previous perihelion passage, and then flared for a few hours, long enough to be seen by Barnard and Brooks. Barnard found another comet in 1886, but Brooks found three within 4 weeks – on April 27, May 1, and May 22. Barnard followed with a discovery on October 4, and on February 16 and May 12, 1887. To close this prolific decade, Barnard found a comet early in 1888 that did not reach perihelion until a year later. Cruising slowly around the Sun, this comet was visible for almost 1000 days. That year Barnard made the transition from amateur to professional by taking a position at the University of California's Lick Observatory, near San Francisco. Although Barnard was now employed with a decent salary, he was still able to use some of his time to continue his search program. His life there was an enjoyable one, at least until the morning of March 8, 1891, when he saw this headline in the San Francisco *Examiner*'s human interest section:

ALMOST HUMAN INTELLECT

An Astronomical Machine That Discovers Comets All By Itself
The Meteor Gets in Range, Electricity Does the Rest.

As Barnard read on with disbelief, he learned of his own fictitious invention of a telescope that would search the sky by itself, making use of that curious newly found property of selenium to react to light.[9] The telescope's selenium device would examine the spectrum of everything. "Stars, nebulae and clusters innumerable crowd into the field with every advance of the clock, but the telescope gives no sign of their presence," the article read. However, should any object showing the "three bright hydrocarbon bands" of comet light appear, the selenium would close a circuit and set off an alarm in Barnard's bedroom. The sleepy Barnard would then rush upstairs to the telescope, confirm that a new comet was indeed there, and return to bed.[10]

Enraged by what was obviously a hoax, and probably also upset that the article portrayed him as needing a lot of sleep (he was never known to sleep much) Barnard wrote angry letter after angry letter. The perpetrator had prepared for this protest by warning the *Examiner* that the reclusive Barnard would undoubtedly deny the story, and the paper refused to print his denials. Even Barnard's colleague, Lewis Swift, who had codiscovered the Perseid meteor comet in 1862, wrote to Barnard asking for all the details. "It takes my breath away, and makes my hair stand straight towards the zenith to think of it," Swift wrote, "Although the article appears somewhat fishy, I am inclined to think it is still another of the marvelous inventions of the 19th century."[9]

Finally, after almost 2 years of embarrassment, and with little support from the Lick Observatory senior staff, Barnard had the chance to prove that he and the newspaper had been deceived. In its issue of February 5, 1893, the *Examiner* regretted "the annoyance which it caused this eminent scientist by printing some time ago an account of a highly ingenious, but non-existent, machine for scanning the skies and catching wandering comets on the photographic plate." The editorial wished him "all the new moons and comets that may be necessary to his happiness."[9] It was even accompanied by an article by Barnard himself, on "How to Find Comets."

As Barnard's friend Heber Curtis wrote years later, "Even as he told me the story, ten years after the event, he was able to summon only a rather wan and rueful smile."[11] Barnard never figured out who was behind the hoax, although the *Examiner* admitted that Charles Hill, who had worked at the Lick Observatory in 1889, had something to do with it. The fact that his own senior colleagues at the Lick did not seem to take his agitation very seriously led Barnard to believe that one of them was behind the hoax. "I never was sure," he admitted to Curtis, "but I have always suspected Keeler."[11] Spectroscopist James Edward Keeler, according to Curtis, "was always ready to laugh at other astronomers, or at himself, if need be."[11] One possible reason for the hoax, if Keeler really did it, was as a prank designed to improve morale at the Lick at Barnard's expense.[12] Barnard himself left the observatory in 1895, accepting a position at Yerkes Observatory at Williams Bay, Wisconsin, where he stayed until the end of his life.

NOTES

1. Norman H. MacKenzie, ed., *The Poetical Works of Gerard Manley Hopkins* (Oxford: Clarendon Press, 1990), 40.
2. M. Proctor and A. C. D. Crommelin, *Comets: Their Nature, Origin, and Place in the Science of Astronomy* (London: The Technical Press, 1937), 112.
3. Ibid., 117–118.
4. Joseph Ashbrook, *The Astronomical Scrapbook: Skywatchers, Pioneers, and Seekers in Astronomy* (Cambridge, MA: Sky Publishing, 1984), 46.
5. Many of the interesting details about Horace Tuttle's life come from an unpublished study *circa* 1980, *H. P. Tuttle: Cometseeker* by Richard E. Schmidt, U.S. Naval Observatory.
6. The basis for this story comes from an unpublished memoir from Richard E. Schmidt, U.S. Naval Observatory.
7. Different parts of this quote appear in these two excellent Mary Proctor books on comets: M. Proctor, *The Romance of Comets* (New York: Harper and Brothers, 1926), 27; M. Proctor and A. C. D. Crommelin, *Comets: Their Nature, Origin, and Place in the Science of Astronomy* (London: The Technical Press, 1937), 154–155.
8. See J. Bortle, "Comet Digest," *Sky & Telescope*, 64, (1982), 294.
9. H. B. Curtis, "The Comet-Seeker Hoax," *Popular Astronomy*, 46 (1938), 70.
10. At the Arthur J. Dyer Observatory of Vanderbilt University, Tennessee, Robert Hardie noted in "The Story of the Early Life of E. E. Barnard" that the university named one of its dormitories after the great comet finder, who was known not to need much sleep.
11. Curtis, 75.
12. J. Lankford, "E. E. Barnard and the Comet-Seeker Hoax of 1891," *Sky & Telescope*, 57 (1979), 420–422.

5

Comet searching in the twentieth century

Hung be the heavens with black, yield day to night!
Comets, importing change of times and states,
Brandish your crystal tresses in the sky,

Shakespeare, 1 Henry VI, circa 1590[1]

"Wanted – a perfect husband, one who wants that happiness not of a day, but of a lifetime; who would receive the fullest pleasure in staying home at night talking to me and would be just as wrapped up in me as in his work."

Imagine a comet hunter, whose interest is typically far more in his or her work than in anything else, responding to this 1915 Chicago newspaper advertisement; but John E. Mellish, a comet hunter and telescope maker who observed from Madison, Wisconsin, did exactly that. Having discovered two comets in 1907, and two more in 1915, he was chosen over some 2000 other men to marry one Jessie Wood, of Glencoe, Illinois. For the first few years their marriage seemed a happy one, and Mellish found his fifth comet in 1917 and his sixth in 1923. However, after nearly two decades of marriage, in 1932, Jessie accused her husband of chasing after a 15-year-old girl. Mellish was unceremoniously thrown into Geneva County Jail, and their marriage was over. He admitted to his indiscretion, although his legal plea was not guilty. "Prosecutor Carbary and Circuit Judge John K. Newhall, under whose jurisdiction Mellish is, admitted today they were in a quandary over disposition of the case," a local newspaper article read. "They said a dozen letters from scientists of seven universities had appealed for mercy."

After this strange episode, Mellish relocated to California where he resumed his long career. Although he found no more comets, he built telescopes almost until his death in 1970. He remains one of the champion comet hunters of this century. After I found my third comet in 1987, well-known amateur astronomer Walter Scott Houston reminded me that "when you get to seven, you will pass Mellish."

Starlight nights

The twentieth century was marked by many significant comet discoveries. Zaccheus Daniel used a 6-inch refractor from the Princeton University Observatory to discover three comets, one of which (Comet Daniel 1907 IV) reached second magnitude. Shortly before Mellish's last comet discovery in 1923, Leslie Peltier acquired Daniel's telescope and began searching with it on a farm near the small town of Delphos, Ohio. "It seemed to me," Peltier wrote, "that if ever human attributes would be invested in a thing of metal, wood, and glass, then this ancient instrument now in my keeping must long for one more chance to show what it could do."[2]

On Friday, November 13, 1925, Leslie Peltier discovered his first comet. In his autobiography, *Starlight Nights*, he tells of that awesome night, movingly describing the search through his opened observatory dome:

> Starting at the horizon, I slowly worked upward back and forth in horizontal sweeps across that bounded bit of sky. Sweep by sweep I climbed upward through Corona, pausing ever so briefly as I hailed, in passing, the patterned landmark of the R Coronae field, and on I moved into the northern end of Bootes.
>
> It was just above the peak of that kite-shaped figure of the Herdsman that the steady cross-sweep of my telescope abruptly stopped. A small, round, fuzzy something was in the center of that sea of stars! A closer, calmer look and I was sure just what that something was, for extending downward from it I could dimly see a slender streak that could only be the tail of the comet I had just discovered![3]

The comet was moving southward so quickly that it was difficult to confirm, but Antoni Wilk, a high school teacher in Krakow, Poland, independently discovered it, and thus it was announced a week after Peltier's find. In the winter of 1929/1930, around the same time as Clyde Tombaugh discovered Pluto, Wilk found two more comets. In 1937, Wilk and Peltier shared a second comet discovery, although this new comet was announced in Europe as Comet Wilk before Peltier had the chance to report his find. This was Wilk's final find. Two years later the Nazis invaded Poland, imprisoning teachers and university professors, and Wilk was among those who spent the war in prison. After heavy international protests, some of the prisoners, including Wilk, were released, but it was too late, for he died shortly afterward.

Peltier continued his comet searching: he discovered a total of 12 comets, and in 1965 published his marvelous autobiography, *Starlight Nights*, a book that I often cite as one of the reasons I decided to begin to search for comets.

Figure 5.1. Comet Levy 1988e (C/1988 F1) taken with the 61-inch Kuiper Telescope and International Halley Watch CCD in the Catalina Mountains, north of Tucson, Arizona, by Steve Larson and the author, in March, 1988.

Comets at Harvard

Harvard College Observatory was a busy place in the 1930s, full of enthusiasm as its astronomers were investigating different fields. One of these astronomers was Fred Whipple, who came to solar system astronomy by an interesting route. As a graduate student at the University of California at Berkeley in 1930, he was part of a team that calculated a rough orbit for the newly discovered planet Pluto. After receiving his Ph.D. from Berkeley, Whipple arrived at Harvard College Observatory in 1931. He wanted to study distant galaxies, but Harvard's director, Harlow Shapley, flatly refused to let him. Galaxies, it turned out, were Shapley's field, and he wanted no one else at Harvard to share it. Moving to the other extreme, Whipple chose meteors as his new area of study, and brought to it his considerable expertise with calculating orbits.

As a part of this work, Whipple took charge of Harvard's photographic plate program, which included the inspection with a hand magnifier of around 70000, 8×10-inch, glass negatives. In some 1200 hours of scanning, Whipple discovered six comets, including, in 1933, the comet now known as 36P/Whipple, the 36th comet shown to be periodic. However, Whipple's major contribution to cometary science would not be the comets he discovered, but what he theorized about the nature of comets and meteors. At the end of World War II, he made the interesting discovery that as a meteoric particle, or meteoroid, orbits the Sun, it slows down so much that eventually it falls into the Sun. Unless the solar system had a way of replenishing its supply of meteoroids, all of these tiny objects should have disappeared eons ago. Whipple believed that comets, which were then thought to be giant collections of sand-like material, must be resupplying the solar system with meteoroids. To explain this, Whipple proposed in 1951 that comets must be comprised of ice and meteoric particles, now popularly known as "dirty snowballs."[4]

The great comets of the sixties

With the end of the Second World War comet hunting became an international sport once more. A team of Czechoslovakian observers, including Antonin Mrkos and Ludmilla Pajdusakova, used large $25 \times 100\,$mm binoculars to find an impressive series of 16 comets. By the mid-1950s, Minoru Honda was finding the first of a dozen comets from his site in Japan. His record inspired a generation of Japanese comet hunters, of which two of the best known are Kaoru Ikeya and Tsutomu Seki. Ikeya began searching in 1962 and found his first comet in 1963, and his second a year later. Guitar instructor Tsutomu Seki was also building a noble record of discoveries by 1965, including a comet that reached naked-eye brightness in 1962, and which was independently found by a husband-and-wife team of amateur astronomers, Richard and Helen Lines.

On September 18, 1965, Ikeya was again sweeping in the southern constellation of Hydra when he found an eighth-magnitude comet. Seki, who was comet hunting that morning too, found the same comet 15 minutes later. "The following parabolic orbit," the International Astronomical Union (*IAU*) *Circular* announced some days later, "shows a close resemblance to that of the Great Sun-grazing Comet of 1882 and its family. According to B. G. Marsden, Comet Ikeya–Seki may be as bright as −7 at perihelion."[5]

On the clear morning of Sunday, October 17, I biked a short distance from my Montreal home to Summit Park, a site with a good view to the east. I thought this would be the best place from which to view the comet, and so did a hundred other people. By 5 a.m. the park was jammed, but although most of the sky was clear, clouds low in the southeast prevented us from seeing the comet.

When Comet Ikeya–Seki rounded the Sun on October 21, 1965, it was bright enough to be seen in daylight even though it was virtually touching the Sun. After a long cloudy period, I finally awoke on Friday, October 29 to a clear sky. I biked once again to Summit Park in time to beat the dawn. As I rode up the last steep hill to the summit I could see, sandwiched between two houses, a comet tail looking as if it was rising out of the St. Lawrence River. I reached the Summit lookout for a splendid view of Comet Ikeya–Seki rising in the southeast. I was alone that morning; the crowd from the week before was gone. This first view of the comet was absolutely unforgettable.

Karou Ikeya's story is special. He developed an interest in astronomy at the age of 13, while he was in middle school in Bentenjima. "I loved to watch the stars through a wide field scope with low magnification. This is the reason why I started comet hunting. I thought that I could contribute a little bit to astronomy if I found a comet."[6]

Ikeya has his own special memories of his greatest comet. He recalls that hectic time thus:

> I remember that it was one to two weeks after the discovery that I knew how close Comet Ikeya–Seki would approach to the Sun. There was a speculation that Comet Ikeya–Seki would evaporate and disappear during closest approach to the Sun. After perihelion I knew that the Comet Ikeya–Seki survived the Sun's heat and became a great comet when I saw the comet tail on the side of the Sun late in October 1965. The comet produced a splendid view early in November.[6]

Meanwhile, Ikeya continued his searching. In 1966 he shared a comet with American astronomer Edgar Everhart, and in the closing days of 1967 he and Seki shared a second comet. That was Ikeya's last discovery for 35 years.

In later years more comet hunters took up the flame. In 1973 Lubos Kohoutek, a professional researcher, discovered two faint comets in the same week. The first stayed faint, but the second Comet Kohoutek, far more distant at discovery, looked like it could brighten in January 1974 to magnitude −10. Although Kohoutek was a bright comet, and was studied by the Skylab astronauts as it passed its perihelion, it failed to live up to those lofty expectations. Meanwhile a second comet was found by William A. Bradfield, an Australian amateur comet hunter. This Comet Bradfield was visible using binoculars during the spring of 1974. "I much prefered Bradfield to Kohoutek," Peltier told me in April of 1974." Bradfield quickly racked up discoveries, and by 1992 his total of 16 tied with that of Barnard.

A Canadian luminary

During the 1970s, the comet hunting pendulum was moving northward to Canada, where the amateur astronomer Rolf Meier, inspired by Clyde Tombaugh's discovery of Pluto, had joined the Ottawa Center of the Royal Astronomical Society of Canada. Meier began searching for comets in 1974, using the Center's newly dedicated 16-inch telescope. He suspected that a telescope of this size would be far more effective in finding comets than the canonical 6-inch wide-field instruments that most searchers use. Thus, Meier began his large-telescope comet hunting as an experiment. He discovered his first comet in the spring of 1978 after only 50 hours of searching, and a year and a half later, after only 29 more hours of searching, he found his second. He found his third comet a year after that. All three comets were found in the northwest evening sky, and he had spent only 105 hours of searching for them all. In the fall of 1984, Rolf and his new wife Linda, fresh from their honeymoon, went observing. As Rolf moved the mighty 16-inch telescope through the sky he enjoyed frequently reminding Linda just how much he loved her. "Oh, Linda!" he would say, and a few minutes later add "My Linda!" Then suddenly his romantic tone changed. The next "Oh, Linda!" was quick and urgent, and followed by "I think I've got one!" This fourth Comet Meier seemed like a wedding gift for the couple. Unlike the unhappily ended marriage of John Mellish, the Meiers seemed to enjoy a union truly made in heaven.

NOTES

1. *1 Henry VI*, 1.1.1–3.
2. Leslie C. Peltier, *Starlight Nights: The Adventures of a Star-Gazer* (New York: Harper & Row, 1965), 127.
3. Ibid., 134.
4. F. L. Whipple, *The Mystery of Comets* (Washington: Smithsonian Institution Press, 1985), 145–147. Two seminal papers discussed Whipple's comet model. The first, "A Comet Model. I. The Acceleration of Comet Encke," *Astrophysical Journal*, 111 (1950), 375–394, explains how the shrinking orbit of Periodic Comet Encke is interpreted if the structure

of its nucleus consists of meteoric material embedded in ice which sublimates to gases. The freed material, which rushes out of the comet with some force, can accelerate the comet. The second paper, "Physical Relations for Comets and Meteors," *Astrophysical Journal*, 113 (1951), 464–474, discusses the implications of this model for comets generally.

5. *IAU Circular* 1925, 1 October, 1965. Also, B. G. Marsden, personal communication, January 2, 1993.

6. Kaoru Ikeya to David Levy, March 15, 2002. Original translated by Shigeru Hayashi.

6

How I search for comets

By being seldom seen, I could not stir
But, like a comet, I was wond'red at;

Shakespeare, 1 Henry IV[1]

Comet hunting demands patience, a good telescope with a wide field of view used under a dark sky, and patience. "In some cases, a comet hunter is not overly focused until he or she finds the first comet," adds comet scientist Daniel Green. "But finding that first comet tends to focus the comet hunter more keenly, and subsequent finds usually occur in much shorter periods of time."[2] Until recently I thought that those words were true, especially in my case. Between 1984 and 1994 I found 21 comets, but although I have not stopped or reduced my comet search program, since 1994 I have not found a single one.

When my comet hunting program began on the night of December 18, 1965, I did not list the actual finding of a comet as the program's primary aim. In the program log that night I wrote instead that I hoped:

(1) To become very familiar with the sky through searching for comets and/or novae.
(2) To discover either a comet or a nova.
(3) To learn as much as possible about comets and/or novae through a research program.[3]

I learned a lot about comet hunting in the months after that chilly December night. The first breakthrough came the following summer, when under the dark sky of the Adirondack Science Camp I was able to spot faint galaxies whose surface brightness was less than the background brightness of my light-polluted sky at home. This meant that my search for comets was likely to be more successful if I could find a dark sky.

Despite this condition, I kept on searching whenever and wherever I could. After 357 hours and 18 minutes of searching, I made an independent discovery of an eighth magnitude comet, Honda 1968c, from a site deep within the city of Montreal and on a night close to full Moon. Even though the comet had been known for several months, knowing that these were the worst possible circumstances under which to find a comet, I was heartened by this incident. Two years later, after a grand total of 458 hours and 32 minutes of searching, I found a second comet in this way, but this time the sky was clear, moonless, and dark. It turned out to be a comet that a Japanese amateur, Osamu Abe, had discovered a few months earlier.

Moving to a dark sky

Having found no comets by the spring of 1979, I decided that year to relocate to the Arizona desert southeast of Tucson. Even so, it was not until April, 1983, after 834 hours of searching, that I found yet another known comet – periodic Comet Tempel 2. At the end of November that year I found Comet Hartley–IRAS only a few days after its discovery! After 863 hours, I was hoping my time was near.

The afternoon of Tuesday, November 13, 1984, was cloudy, but a clearing sky at sunset made me cut short a dinner date and rush home to begin my comet search. As was my usual practice I moved Miranda, my 16-inch reflector telescope, from field to field. After about an hour of searching I came across the open star cluster NGC 6009; but just to the south was a diffuse object. The sight of cluster and fuzzy object was so striking that I wondered why I had never seen it before. My *Skalnate Pleso Atlas of the Heavens*[4] confirmed the presence of the star cluster, but not the fuzzy object. I began to get really excited. I sketched the cluster, a few field stars, and the fuzzy patch. When I checked the field a quarter hour later, I was sure that the object was moving very slowly in the direction of the cluster to the north.

It was a comet, but was it already known? I spoke over the telephone with Brian Skiff, an observer at the Lowell Observatory some 300 miles away in Flagstaff. After he had checked the positions of all known comets, Brian said "You'd better send a telegram. You've got a comet." After 917 hours and 28 minutes, spread out over 19 years, my search was over.

Building on success

Or was it? I had a feeling of absolute satisfaction and relief as I sent a telegram to Brian Marsden, Director of the International Astronomical Union's Central Bureau for Astronomical Telegrams, with the news of the new comet. The following evening he spoke with me over the telephone. It turned

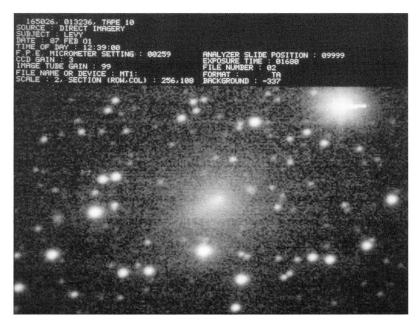

```
  165026, 013236, TAPE 10
SOURCE   DIRECT IMAGERY
SUBJECT : LEVY
DATE : 87 FEB 01
TIME OF DAY : 12:39:00
F.P.E. MICROMETER SETTING : 00259      ANALYZER SLIDE POSITION : 09999
CCD GAIN : 3                           EXPOSURE TIME : 01680
IMAGE TUBE GAIN : 99                   FILE NUMBER : 02
FILE NAME OR DEVICE : MT1:             FORMAT :        TA
SCALE : 2, SECTION (ROW,COL) : 256,108 BACKGROUND : -337
```

Figure 6.1. Comet Levy 1987a (C/1987 A1) photographed by Jim Scotti with the Spacewatch camera at Kitt Peak in Arizona, in January, 1987. Courtesy James V. Scotti.

out that Michael Rudenko, an amateur astronomer and comet hunter from Massachusetts, had just discovered the same comet. Thus the new object became known as Comet Levy–Rudenko, in continuation of a tradition of naming comets after their discoverers that dates back to the mid-eighteenth century time of Charles Messier.

Halley's comet came and went between my first and second comet finds. Early in January, 1987, I found my second comet as a faint fuzzy visitor in a predawn chill. The sky was already starting to cloud over in anticipation of an approaching storm, and as I plotted the comet's position on my *Atlas Eclipticalis* heavy rain began to fall. The comet was faint, and never became brighter as it headed out of the solar system on its long, slightly hyperbolic path.

I did not have to wait long for my next find. On October 12, 1987, while testing a new observing site atop the roof of my home, I discovered my third comet only 107 observing hours after the second. Nor did I have to wait long for a fourth. On March 19, 1988, I found a comet (Comet Levy 1988e) that began a strange story and a wonderful friendship. That story began early in March, 1988, at a conference in Tucson called "Asteroids II". It was at that conference that I met Gene and Carolyn Shoemaker, two geologists who had begun their own comet search program a few years earlier. While my program's only scientific purpose was to learn as much as possible about comets, theirs had a more specific design: To gather statistics about the numbers of comets and asteroids that could pose a threat to Earth. I was fascinated by this new area of study and by the importance that they had given to the role comets have played in the history of Earth.

Two weeks later, on the morning of March 19, I discovered Comet Levy 1988e while the Shoemakers happened to be observing at Palomar. At the end of one of their long and productive nights, they positioned the telescope far to the east and took a brief exposure of the field that contained the new comet (as reported in the latest *IAU Circular*). Despite the brightening dawn sky they got a good image and submitted the first accurate positions of my comet.

The next month Gene and Carolyn, now with colleague Henry Holt, included the new Comet Levy on their list of fields of sky to photograph. They set up their list graphically on a sheet of paper on which nickel-sized circles outline the observing fields around the sky, but the comet was north and east of where they usually photograph, so Gene placed an extra circle at the top of his diagram. On the morning of May 13, 1988, the telescope was pointed at the position indicated by the extra circle. The following evening, Carolyn placed the films showing Comet Levy on her stereomicroscope and quickly found what she thought was my comet. Gene looked also, but he wondered why the comet was so far from the field's center. Carolyn then tried to measure the comet's position relative to the surrounding stars. To her frustration, the stars on the field did not match any of those on the field they were supposed to be photographing. They had photographed, it turned out, the field represented by the position of the circle on the diagram, not the Comet Levy field; by attempting to observe a known comet, they had discovered a new one.

As is usual for the period immediately following a discovery, astronomers around the world strive to obtain precise positions of the comet and submit them to the Minor Planet Center. Using these positions of the new comet, named Shoemaker–Holt 1988g, Conrad Bardwell of the Central Bureau for Astronomical Telegrams made a discovery of his own. The orbits of Comet Levy and Comet Shoemaker–Holt, he noticed, were almost identical in every respect except that Comet Shoemaker–Holt arrived at its closest point to the Sun, or perihelion, some 3 months after Comet Levy's closest approach to the Sun. This was the first case of a pair of related long-period comets being discovered independently. The two comets were one some 12000 years ago and, for some reason, split apart.[5] They are continuing their separate journeys around the Sun, moving away from each other as they go. The two comets will probably be years apart when they next return.

Comets and a spacecraft

On August 26, 1989, the spacecraft Voyager 2 was making its dramatic flyby of Neptune and its moon Triton. On that evening, fresh from completing an observing period at the Catalina 61-inch telescope, I was free to begin some comet hunting with Miranda. Even though August is known for its poor weather in southern Arizona, that particular night was clear. Since the images of Triton were being broadcast live on the American Public Broadcasting System that night, I alternated between watching each new photograph of distant Triton appear on my television set, and heading outdoors to search for comets. After an hour and a quarter of intermittent searching, my views of Triton were completely interrupted by my discovery of a new comet. I shared this discovery with Kiyomi Okazaki and my friend Michael Rudenko, who was conducting his own search program from Massachusetts. This new comet slowly brightened over the next few months, reaching near-naked eye visibility for a short time.

A few months later, the world was waiting for a bright comet. Found by Rodney Austin at the end of 1989, from New Zealand, Comet Austin $1989c_1$ promised to become as bright as Jupiter as it moved northward toward the Sun. In May 1990, I was observing with Steve Larson from the Lunar and

Figure 6.2. Comet Levy 1990c (C/1990 K1) photographed by the author with the 18-inch Schmidt camera on Palomar Mountain, California, in August, 1990.

Planetary Laboratory, at the large 61-inch telescope in the Catalina Mountains north of Tucson. We had planned to observe the comet in different filters, and to obtain spectroscopic images using the International Halley Watch CCD (electronic charge-coupled device) system. By the fifth night of our run, it was clear that Comet Austin would not perform as we had hoped. "David," Steve said early in the morning of May 19, "Comet Austin is a one-person job. Tomorrow night you should stay home and find us a bright comet!"

Although the sky the following morning was lit by a last quarter Moon and clouds, I decided to accommodate Steve by hunting earlier than usual. I opened the roof and began hunting around 2:30 a.m. I decided I would continue until moonrise, but the clouds were interfering even with that plan. However, as the Moon rose an hour later, the clouds dissipated and I kept on searching. Swinging the telescope north to keep as far from the Moon as possible, I hunted through the constellation of Andromeda, past the bright star Alpheratz, and then I moved on into Pegasus. Not far from Alpheratz, a soft fuzzy patch of light entered the field of view. Since I knew this part of the sky very well, I immediately thought this must be a comet. Yet over the next hour it did not appear to move relative to the surrounding stars. A check of the Palomar Sky Survey showed nothing in that position. Twenty-four hours later the comet had crept along barely an eighth of a degree, by far the slowest motion I had ever seen for a comet at discovery. Announced as Comet Levy 1990c, this new visitor was far from Earth and heading toward us (see Plates I and III).

By early July the new Comet Levy was picking up speed and brightness as it approached both Sun and Earth. In late July, while visiting my family, I went south of town to show the comet to a reporter for the Montreal *Gazette*. The sky had been cloudy for a few days, so I was not sure what the comet would look like, but when we arrived at the site, I looked up and saw the comet clearly visible to the naked eye, and through Minerva, my 6-inch-diameter portable telescope, it was a marvelous sight. By the middle of August it cut a striking path right near the summer Milky Way. The comet remained fairly bright well into the early months of 1991. In February 1991 it crossed the plane of Earth's orbit and for 2 or 3 weeks sported a beautiful anti-tail (see Chapter 15 for a description and photograph). In May 1991, I observed my comet on the first anniversary of its discovery. Very much faded and hardly visible, its long show was finally over.

Comet hunting via photography

In the fall of 1989 I began a new kind of comet searching. My 1988 meeting with the Shoemakers led to me becoming a partner in their comet and

Figure 6.3. Comet Levy 1990c (C/1990 K1) photographed by the author through the 18-inch Schmidt camera on Palomar Mountain, California, in August, 1990. A satellite, tumbling over in space, is crossing the field.

asteroid search, which took place for a week each month at the 18-inch telescope at Palomar. Our procedure involved taking a film covering an area of sky, then taking a second of a different area, a third, and then a fourth. We would then repeat these exposures so that each area was exposed twice, the different exposures being about 45 minutes apart. During a typical night we might take five or six sets of film, and over a full week we would cover a considerable amount of sky. In November 1990, the first Shoemaker–Levy (1) comet shed a few photons of light onto our films, but even though we did not know it at the time, this did not turn out to be our first joint photographic discovery. The previous month we had found an asteroid, designated 1990 UL3, with a most unusual orbit that took it out toward Jupiter's distance from the Sun and back again in a period of about 6 years. This kind of orbit seemed more like a comet than an asteroid orbit, and Brian Marsden, then director of the Minor Planet Center, asked Carolyn to check the discovery images for any sign of cometary activity. She reported back that there did not seem to be any fuzzy appearance to this object. However, our discovery telescope at Palomar was only a wide-field 18-inch Schmidt camera. I was scheduled to observe using the much larger 61-inch telescope and CCD system, of the Lunar and Planetary Laboratory with my colleague Steve Larson, so I suggested to him that if 1990 UL3 were indeed a comet, this telescope might be able to detect the faint glimmer of a coma surrounding the asteroidal-appearing central condensation.

On the evening of December 18, 1990, we began a series of 5-minute exposures of the field containing 1990 UL3. When the first image appeared on our computer screen we saw just a field full of stars, as well as faint lines that showed the weak areas of the camera's CCD chip. Then we calibrated the image using bias and flat-field frames, a procedure designed to increase the clarity and sensitivity of the image. Once we did that, it was easy to see that near the bottom of the field, one of those "stars" had a tail! Asteroid 1990 UL3 was now Periodic Comet Shoemaker–Levy 2.

1991: A cometary *annus mirabilis*

The year 1991 was so successful as a year of comet discovery that it almost spoiled us into thinking that we would find at least one new comet each month we observed at Palomar. At our first run that January, on the night of January 21, the sky finally cleared to a cold, windy night. On the films we exposed that night was Comet Shoemaker–Levy, 1991d. On the first night of our February observing run, February 6, we took films that contained periodic Comet Shoemaker–Levy 3. Just two nights later, on February 8, our films captured periodic Comet Shoemaker–Levy 4.

At our next observing run, we found a comet on films taken on March 11, which turned out to be a known periodic comet, discovered on its last return by Malcolm Hartley, but far from its predicted position. On the morning of June 10, I awoke to search the morning sky, but found that it was cloudy. Nevertheless, I opened the sliding roof on my observatory. Soon the first signs of dawn were appearing, and a clearing formed in the eastern sky. I began sweeping in Aries where, after only 1 minute, I spotted a bright hazy spot that I quickly identified as Messier 74, one of the faintest of the objects that momentarily fooled Charles Messier. The sky continued to brighten as I moved the telescope a few fields over. After another minute, I saw another fuzzy spot. Could I have stumbled on Messier 74 a second time? This was not likely: the object was brighter, and the field of stars was clearly different. More importantly, where the galaxy M74 had moderately sharp edges, this object showed the gradual fading at the edge which is more typical of the gas and dust in a comet. This latest Comet Levy was traveling along the ecliptic, so I was not too surprised to learn weeks later that it was a new periodic comet that returns to the vicinity of the solar system every half century. For some reason it had never been picked up earlier, with one possible exception: In 1499, Chinese and Korean observers observed a comet pass from Hercules through Draco, and the Little Dipper (in

Ursa Minor) and Big Dipper (or Plough, in Ursa Major).[6] The orbit of that comet is so similar to that of periodic Comet Levy that it could be the same comet, although positive identification will probably have to wait until the comet returns in around 2041.

One month later I was far from home, in La Paz, Mexico, observing a total eclipse of the Sun with an almost record length of totality approaching 7 minutes. On the night of September 7, in less than 2 hours of comet hunting I independently found two already-known comets, periodic comets Hartley 2 and Wirtanen. Back at Palomar that autumn, we discovered Comet Shoemaker–Levy 1991z on films taken on October 1 through a sky brightened by dust and acid from the eruption of Mt. Pinatubo. Six nights later we took films that revealed Shoemaker–Levy $1991a_1$. Our next observing run began in early November. We found periodic Comet Shoemaker–Levy 6 on November 7. The comet was bright enough to observe visually through Minerva, the 6-inch f/4 reflector that always accompanied me on these observing runs. Finally, as we observed one terribly windy night, we tried to keep the telescope pointed east, and out of the wind; but as the night progressed, our search program called for a field far to the northwest, so we had to brave the gale. Although I tried to keep the telescope steady, the thin metal shutters at the top caused it to shake like a sail. The guide star was dancing about so violently in the field of view of the eyepiece that I was unable to keep it centered in order to guide the exposure. Then Gene had an idea. For the next exposure, he stood precariously on the top of an elevating chair, reached to the top of the telescope, grabbed the shutters, and held them. I swung the telescope into place. As Gene's chair rose several feet; he stood on it to reach the top of the telescope. I peered through the eyepiece at the guide star, which was still dancing about like a firefly. All was ready: I pulled the lever and the shutters slowly opened. As Gene grabbed on to them, the guide star settled down. For the next few minutes the wind howled as Gene held on to the shutters from his unsteady position on the chair, trying to keep the telescope steady and himself from falling. Meanwhile I tried gamely to keep the guide star centered. Carolyn was scanning films downstairs with her stereomicroscope. Curious about the movement of chairs in the dome, she climbed the stairs to see her husband standing precariously atop the chair. "Is this exposure absolutely necessary?" she asked.

The noise from the wind was so loud that we could hardly hear our own words. "What?" Gene hollered. "IS THIS EXPOSURE ABSOLUTELY NECESSARY?!" she repeated.

"YES!" Gene hollered back, his voice still difficult to hear through the pounding wind. We imagined his thought after that: "You never know what's in the field if you don't shoot it."

About 45 minutes later we had to repeat the exposure, but then we turned the dome to the east and continued observing in the more sheltered areas. It turned out that Gene was right to have persevered under those windy conditions, for when Carolyn scanned these films, she found Shoemaker–Levy 7. Thus our *annus mirabilis* of 1991 ended with seven Shoemaker–Levy comets and one Levy comet.

Our discovery rate dropped sharply after 1991. On films taken on April 7, 1992, we found Comet Shoemaker–Levy 8. On March 23, 1993, we took the discovery films for Comet Shoemaker–Levy 9. One more Shoemaker–Levy comet turned up in 1994, and on April 15 of that year, while searching visually, I discovered Comet Takamizawa–Levy, a comet then crossing the tiny constellation of Equuleus. My search still continues, visually, photographically, and with CCDs. Although it has been a very long time since a discovery, I remind myself of the primary aim of the project, to become familiar with the sky through searching for comets. In that sense, I learn something new about the sky every time I open wide the observatory roof and begin a night of searching.

NOTES
1. *1 Henry IV*, 3.2.46–47.
2. Daniel Green, ed., *International Comet Quarterly Guide to Observing Comets* (Cambridge, MA: Smithsonian Astrophysical Observatory, 1997), 61.
3. CN-3 Record Book, 1965. (This is the official comet hunting record that I have kept since 1965.)
4. Antonín Becvár, *Skalnate Pleso Atlas of the Heavens 1950.0* (Cambridge, MA: Sky Publishing, 1949).
5. Brian Marsden, "The Comet Pair 1988e and 1988g," unpublished paper.
6. Donald K. Yeomans, *Comets: A Chronological History of Observation, Science, Myth, and Folklore* (New York: John Wiley, 1991), 410.

7

Searching for comets photographically

But however entrancing it is to wander unchecked through a garden of bright images, are we not enticing your mind from another subject of almost equal importance?

Ernest Bramah Smith, "The Story of Hien and the Chief Examiner"
in *Kai Lung's Golden Hours*, 1922[1]

Edward Emerson Barnard might have found a way to get even with those anonymous souls who had perpetrated their cruel hoax on him (see Chapter 4). In 1892, 18 months after the prank, but still well before the *Examiner* finally retracted their story, Barnard actually did discover a comet in a completely new way. After developing a photographing plate of a region of the sky that included Altair, one of three stars in the summer triangle, he saw a fuzzy trail that he correctly interpreted as the image of the comet moving either from southwest to northeast, or northeast to southwest. The new comet was named periodic Comet Barnard 3. Barnard then exposed a second plate of the same area that confirmed the object was really a comet, and that it was indeed moving toward the northeast.

Three quarters of a century later, much had changed, for by then (the mid-twentieth century) the discovery of comets on photographic plates had become common. "By far the great majority of new comets are now found on the photographic plate," Leslie Peltier wrote me in September 1966, "and this lessens, of course, the amateur's chance of making a find." Peltier was right, and would still be today, only that astronomers are now using electronic CCD images instead of film. However, film still has advantages over the more modern CCD. A single sheet of circular film, cut out of a $2\frac{1}{4}$-inch square sheet, in one of our 8-inch f/1.5 Schmidt cameras, can cover some 78.5 square degrees of sky – the area covered by a small constellation like the Southern Cross – in a single exposure! To get the same coverage with a CCD that covers one degree of sky, you would need 79 exposures.

Both film and CCDs are subject to the common errors of ghost images and artifacts, and these problems were especially common in the earlier years. Around the time of Peltier's letter to me, physicist Glenn Shaw, then a graduate student, saw a perfectly shaped comet on a photograph he had taken the night before atop Mount Wilson. He showed the photograph to Jesse Greenstein and Maarten Schmidt, two of Caltech's most famous astronomers. They both thought that the image could be real, and set up time on Palomar's 48-inch Schmidt camera, the world's largest, to confirm the comet; but the big camera did not record any comet. Shaw had recorded one of the earliest of many examples of a plate defect or a static discharge, either of which can resemble a well-developed cometary image.[2] Ghost images, where the reflection of a bright star on the opposite side of a film resembles a moving comet, and other artifacts plague photographic comet hunters all the time. Ghosts have visited our program many times. One day Wendee detected two perfectly shaped cometary structures on two films, indicating that a comet that moved an appropriate distance between the times that the two pictures were taken. However, the images did not look exactly alike – a giveaway that both were static discharges.

Despite the proliferation of ghost images from bright stars and static discharges, photography has clearly proven itself as a comet detector.

Preparing film for exposure

The key to effective use of film for comet hunting is to find the finest grain film possible as well as the fastest film possible. The old Tri-X was very fast, ISO speed 400, but its coarse grains make it hard to locate any comets. Perhaps the finest grain black and white film available is Kodak's Technical Pan 2415 or its thicker-based cousin 4415. However, this film, meant for copying, is very slow, requiring long exposures.

There is a way, however, to increase the film's sensitivity. The technique, called hypersensitization, uses an oven set to 65 degrees C, and a flow of "forming gas" usually consisting of 92 percent nitrogen and 8 percent hydrogen. Hydrogen, the active ingredient, flows over the hot film for 6 hours, drying it and increasing its sensitivity. Some films react more favorably to this "hypering" than others; Tech Pan is one of the better ones, increasing its sensitivity by a factor of 9.

Scanning with a stereomicroscope

About a century ago the German asteroid hunter Max Wolf thought of the idea of using a stereomicroscope to compare identical images of stars in a search for moving objects. He did not use the device effectively, according to one story, because one of his eyes was weak. Instead, he developed another device called a blink comparator in which the first image, then the second, would successively come into view. In this way he would use his better eye to view the images. A moving object would appear and disappear as one plate cycled through, then appear and disappear again in a slightly different position on the other. In 1930, Clyde Tombaugh used a blink comparator to discover the planet Pluto; in later years he used the device to find many asteroids, one comet, a new variable star, six star clusters, and a supercluster of galaxies.

The stereomicroscope came back into vogue in the 1980s through the insight of astrogeologist Gene Shoemaker, who transferred the technology of looking at two aereal photographs of the same region of Earth taken from very slightly different angles to give a three-dimensional impression of depth. Using such an instrument on astrophotographs, both images are examined *simultaneously*. If a comet or asteroid happens to be in the field, it will appear to float above the background of stars, or to sink below that background. Using a stereomicroscope allowed Gene's wife Carolyn to examine two pictures of the same region of Earth taken from very slightly different angles; the resulting pair gave a three-dimensional impression of depth.

The Shoemaker–Levy Double Cometograph

In Chapter 6, I described the Palomar photographic experience I shared with Gene and Carolyn Shoemaker.[3] After we ended our Palomar Asteroid and Comet Survey in 1994, we set up a new program at Jarnac using a series of small Schmidt camera/telescopes. Seven years later, we are continuing this program even though we have yet to make a comet discovery with it. The photographic mode of the program consists of the following components:

(1) Ophelia is a set of twin 18-inch f/1.5 Schmidt cameras, manufactured by Celestron International but improved and refitted by Epoch Instruments. Each camera is fitted with a special Vehrenberg film holder allowing a coverage of 10 degrees of sky in a circular field. Although the two cameras share a single mount, Ophelia 2 sees a field of sky centered about 10 degrees north of what Ophelia 1 sees.

(2) Obadiah is a 12-inch f/2.2 Schmidt camera, manufactured by Meade Instruments Corporation and fitted with a film holder designed to cover 4.5 × 5.5 degrees of sky in a rectangular field.

(3) Telescope encoders and position readouts are used for both cameras. Obadiah uses its own LX-200 software; Ophelia employs the NGC-MAX

device, manufactured by JMI. Sky Commander is a similar device that can provide positions.

(4) We use an old copy of *Norton's Star Atlas* (given to me by my Dad, actually, back in 1964), with 10-degree fields (100 square degrees) marked off. Fields 1–175 are the nova search fields straddling the Milky Way and were created by the American Association of Variable Star Observers; fields 176–428 , covering the rest of the sky, are comet search fields created by Jim Low of the Royal Astronomical Society of Canada.

(5) We have a series of recording sheets that cover the required information for recording each exposure's field number, listed position in right ascension (RA) and declination (Dec), and the magnitude of the star we guide on. Because the positions of the double cometograph mount are not as accurate as we would like, we record actual positional readouts in two additional columns that show actual RA and Dec, after the star is centered in the guiding eyepiece. The form also has a comments section in which we record the appearance of the field around the guide star (e.g., bright star at 3:00 a.m., faint double at 7:30 a.m.), so that we can further confirm that we are guiding on the same star when we return to take the second picture of the field. The Meade camera LX-200 electronics are so accurate, always repositioning the guide star to the center of our high power field, that these backup steps are not necessary.

(6) We have a series of finger lights painted red. Wendee found these very small flashlights which attach around a finger with a velcro strip. We need lights often during each session, particularly before and after each exposure; however, the illumination cannot be underfoot in case they interfere with operations such as changing film.

(7) A lightproof film changing closet is a must. We needed a convenient space for changing films that would allow for quick film changes (i.e., avoiding a dash inside to a darkroom every 8 minutes) but which would protect films from light. Wendee designed a wooden closet with hinged doors, and attached a dark towel with small C-clamps from the top of the closet, allowing for freedom of hand movement inside while preserving our protection against light.

(8) Each telescope has a timer set for the lengths of its exposures: 8 minutes for Ophelia; 15 minutes for Obadiah, which has a longer focal length.

A search strategy

The most important part of comet searching is to develop a strategy. With film, the plan should be designed to cover as much area of sky as possible without covering areas that are already well searched by the professional

Figure 7.1. Comet Ikeya–Zhang (C/2002 C1) photographed by Wendee Levy, Carolyn Shoemaker, and the author using a 12-inch Schmidt camera at the Jarnac Observatory (Vail), Arizona, in February, 2002.

surveys. Our current program searches the sky within 90 degrees of the Sun, and south of −30 degrees.

With our 8-inch f/1.5 Schmidts, an 8-minute exposure is usually enough. With other setups, to determine how long an exposure you should take, you just try a series of different times with the film and telescope chosen. If the sky background is visible as a general darkening of the film that has been exposed to the sky, then the exposure time should be sufficient.

After we take an exposure of a field, we then take a second field, then a third, a fourth, and sometimes a fifth. When the series is over, we then repeat all the exposures. Ideally, between 45 minutes and an hour should separate each pair of exposures. When the night's exposures have been completed, they are photofinished using a high contrast developer like Kodak's D-19, then fixed, and finally washed for 30 minutes and allowed to dry.

Observing procedure notes

Running three photographic telescopes (not to mention two additional automated CCD cameras that I will feature in the next chapter), all working simultaneously, requires some organization. Wendee is in charge of the night's program, deciding how many fields we will cover with each telescope. She also loads and unloads all the films into their proper filmholders, a formidable task when all three cameras are operating. Once I know how many fields to take, I choose the particular fields based on their position in the western sky – they must not be so high that the major surveys are covering them, nor can they be so low that I am uncertain that I can photograph them twice during the evening with sufficient time elapsing to allow for us to detect motion of any comets or asteroids.

Thus we begin each night by choosing which fields we want to photograph, using *Norton's Star Atlas* and our view of the darkening sky. Each telescope is set up and aligned on a star. Carolyn brings her vast experience on large Schmidt cameras, like the 18-inch camera at Palomar, to run Obadiah, while I usually run Ophelia.

When the film is loaded into the camera, the guide star position is read and the telescope pointed. Once the guide star is centered, Wendee *gently* removes the cover from the camera's glass corrector plate. She then holds the cover about 2 cm away from the camera while Carolyn or I make final adjustments to the position of the guide star. (As an aside, if a guide star happens to be spectral class G, it is usually so noted. It is nice to imagine, while staring at a distant star for 8 minutes, that someone living on a planet circling that star might be looking back.)

After the exposure has started, Wendee records the time to the nearest 5 seconds, and the observer informs her of the current position of the telescope. When it is time to repeat the exposure (from 40 to 60 minutes later), these actual positions are the ones used.

On a typical evening, we will take four or five pairs of photographs using both Ophelia cameras, for a total of 16 or 20 pictures. Meanwhile, we take three or four fields with Obadiah, for a total of six to eight photographs. By the time we are finished, we have the option of proceeding to a second set, but we do not often go on after our first set is done. By now the fields that are available for photographing are the ones that are already covered by the professional surveys. It is my custom to concentrate on the photographic searches during the evening hours, reserving the time before dawn for visual searching.

A focusing tale of woe

The procedure I have just described sounds so simple and pleasant, but it is the result of years of fine tuning experience conducted both with this program and with the earlier survey conducted by the Shoemakers at Palomar with the 18-inch Schmidt camera. The most aggravating problem we ever had was focusing the cameras and keeping them in focus. At Palomar, focusing was done when needed, usually at the start of the observing run, by this process: We loaded a film into the camera and made a 10-second exposure of a fairly bright star. We then moved the telescope slightly in right ascension, changed the focus and took a second exposure on the same film. We repeated this process several times, recording the focus numbers. After the film was developed and fixed, we examined the line of star images under a microscope, choosing the clearest image and setting the telescope at the correct number. As the temperature dropped during the night, we adjusted the focus according to a table.

Ophelia and Obadiah use invar rods, which do not change their lengths with temperature, to support the film holder and keep it at a fixed distance from the primary mirror. However, Ophelia's design makes it almost impossible to focus except in a laboratory. Bob Goff, an accomplished telescope optician, succeeded in focusing one of the cameras, Ophelia 1, without difficulty. Ophelia 2, however, seemed to elude his laboratory efforts. Bob finally suggested that the only way to focus the camera was to do so under the stars. We began one clear evening by taking, developing, and examining a single photograph. Then Bob used a wrench to adjust each of the three pairs of locking bolts for the invar rods the tiniest fraction of a turn, and we repeated the exposure. We improved the focus after a few tries, and continued the project some nights later.

Focusing Ophelia 2 almost became an end unto itself as we continued the slow and painstaking process, which also had some unintended consequences. One evening, while facing the telescope pointed to the zenith and standing atop a ladder with wrench in hand, this world-renowned telescopist started answering my questions using a perfect imitation of Donald Duck. "How close are we to focus?" I asked. "Closer and closer, I hope!" Bob quacked. Another evening, while hurrying to meet Bob and drive him down to our house, I got a speeding ticket. To avoid a fine I had to take an all-day course in driving safety.

Wendee, who audited the course with me, noted that the instructor had a hard time believing that the reason I got the ticket was that I was rushing to get home by nightfall to start focusing a camera.

Bob Goff was a very dear friend, and we were saddened by his death just before the end of 2001. By this time we easily could have abandoned Ophelia 2 altogether – after all we did have two well-focused instruments; but we wanted to finish the job partly out of respect for Bob. So not long after his memorial service, Dean Koenig, another good friend and telescope expert, resumed the focus effort. This time we worked all night long, taking one film after another. To save time I moved the photofinishing chemicals out into the observatory, and we kept the developing and fixing times to the absolute minimum needed to show images. Still, it was a painful process. As the night went on the temperature dropped and with it, the efficiency of the chemicals. We also realized that the fainter stars gave us far more information about the state of focus than did the brighter ones. A bright star might appear focused, while the faintest ones around it might be slightly fuzzy. So we reluctantly decided to take almost full-length exposures, then use the normal time-and-temperature method to develop them. With 8-minute exposures, 10 minutes of developing time, 5 minutes of fixing, plus time for examing the films, and more time to adjust the focus, we were lucky to get three shots per hour.

We repeated this process over five or six nights spread out over several months. During one excruciating session, 12803AN on April 16, 2002, we thought we were there. Most of the shots we had taken moved the telescope closer to focus, but one side might be sharp while images on the opposite side, near one of the invar rods, would appear elongated. Dean finally got the elongations to a minimum. We could have stopped there but decided to see if we could improve the focus further. The next adjustment turned out to be in the wrong direction; when we turned the nuts backward another side of the shot began to go out of focus, and by the end of the night our films sported doughnut-shaped star images, the signature of a seriously out-of-focus telescope. "We went through perfection," Dean said over the telephone to his wife Donna, "and came out on the other side!" We stopped for that night, with an ever-optimistic Dean saying, "Don't worry, we'll get it yet! We got a lot of information tonight about how the telescope behaves. We'll get it." Dean was right of course; with each turn of the nuts supporting one of the telescope's three invar rods, and seeing the result, we were getting to know our telescope better and better. We ended the night with a collection of many examples of poorly focused and aligned star images.

At last, the great night came: April 25, Session *12820EM2. Dean made change after incremental change, carefully making notes of every step he took.

Each change brought the telescope closer to focus. Late in the night, while dangerously close to focus, we had to make a fateful decision. Should we continue trying, and risk losing what we had accomplished? We decided to continue, and this time the effort happily paid off. We ended the night with a telescope focused better than it had ever been before.

What lessons did we learn from this 3-year long exercise? The most important one is that it is not enough just to go out and buy a telescope or camera; it is important to get to know it. Our case was an extreme one, thanks to a poor original design of the attachments for the invar rods, resulting in it taking years (actually many hours and all-night sessions) to adjust the telescope so that it would produce consistently acceptable results; but somehow, when we examine a picture taken by this camera, we appreciate it all the more. Beautiful as the night sky is, we have a greater appreciation of it when we understand how the telescope works to capture photons of starlight, and then to record those photons on photographic emulsion.

NOTES

1. Ernest Bramah Smith, *Kai Lung's Golden Hours*, 1922 (New York: Crown, 1967).
2. *Let's Talk Stars* radio show, KTKT Tucson, Arizona, May 7, 2002.
3. Carolyn S. Shoemaker "Twelve years on the Palomar 18-inch Schmidt," *Journal of the Royal Astronomical Society of Canada*, 90:1 (1996), 18–41.

8

Searching for comets with CCDs

Little One! Oh, Little One!
I am searching everywhere!

James Stephens (1882–1950), The Snare[1]

CCDs are very much more sensitive to light than photographic film or plates, but because wide-field versions are very expensive, using one to observe or discover a comet is not widely practiced by amateurs. However, the numbers of amateur astronomers with this equipment have increased in recent years, and as prices for CCDs continue to drop, the technology will become more popular. This chapter offers an overview of what equipment, software, and other effort is needed to launch a CCD program for comets.

The CCD

The charge-coupled device or CCD, its housing, and its electronics are by far the most expensive part of what you will need, more so even than your telescope. If you are interested only in observing comets that have already been discovered, the narrow-to-moderate field of view that the chip will be called upon to cover means you can invest in a relatively small chip. If you are using a CCD to hunt for comets, then you should get the largest chip that you can afford.

There are a variety of CCD systems on the market. Pricing is competitive, but generally the larger the chip, the higher its price. Do not ignore an opportunity to buy a used CCD system; however, learning to use such a system is not easy, and after buying one and trying it out, some people feel that they are in over their heads and are willing to sell it at a big discount.

The computer and its ports

The faster the computer, the more effectively a CCD system will work. It is not uncommon for a CCD image to download into the computer over 5 minutes or more, which is far too slow for an effective survey program. The speed of download can depend on which port is used. The computer's serial port is usually the slowest. The SCSI port can be the fastest but in my experience it is also the most sensitive; if I even touch the mouse during a 5-second download, the image might be lost. Other systems use the printer (or parallel) port, or one of the other com ports.

Finally, if your computer screen is out in the observatory with you, it should be darkened with a large red filter so that your eyes will not lose their dark adaptation.

The software

A well-written, user-friendly system that runs the CCD is at least as important as the CCD itself. A good program will help focus the camera, take pictures, allow you to view the result at different contrast levels, and do some manipulation of the image. I use MAXIM DL to control my CCDs. Written by Doug George, a visual comet discoverer from Ottawa, Canada, MAXIM can control almost all makes of cameras.

For comet searching, I use Bob Denny's Astronomer's Control Program (ACP) to control both the CCD and the telescope. ACP wraps its arms, so to speak, around MAXIM DL, which is used to turn on the CCD and set the parameters for calibrating the image; MAXIM continues to work silently in the background. ACP moves the telescope from target to target, and it even measures the position of every image using a software component called PinPoint. Once the observing list for the night is prepared, ACP works with the telescope and CCD, photographing each field three times.

After the session is over, I use Visual PinPoint to scan each trio of photographs. Again, this process is automated; it can be set so that I do not even see the images unless the program has detected what could be an asteroid or comet. Then the three images will blink one after the other, while a circle shows the position of the suspected object. If the suspect turns out to be a real comet or asteroid, the program even records its positions in the format preferred by the Minor Planet Center.

How to connect a CCD to a computer

Connecting a CCD system to a computer and then to a telescope takes time and patience. The first two steps involve attaching the device to your telescope and your computer, and the CCD instruction manual should show how this is done. Unfortunately, some manuals are not written for a person without a lot of computer experience, so if you are like me then you might have to proceed with caution, or at least expect to have to learn as you go. Electronic observing is a whole new experience, and it takes time to get used to it.

In order for the system to work effectively, it must first be cooled to a temperature far below room temperature. The first system I used, a professional system meant for observatories, had to be cooled to -90 degrees Celsius using liquid nitrogen. Most systems meant for amateur use are cooled electrically; this is a far more convenient procedure that works when a system does not need to be cooled as much.

Focusing

Most cameras, and most software, offer a focus mode where photographs of a small area of the chip, centered around a star, are taken and displayed, one after the other in rapid succession. This way, you can adjust the focus quickly until you are satisfied that the tiny image is sharp. Adjust it through perfect focus, then out on the other side of focus a bit, until you are certain that you have the best possible focus. Keep in mind that the small image displayed on the screen might not be the last one the camera took. After making a focus adjustment, wait a few seconds to make sure that you know which image you have used to decide which way to change the focus.

If the atmospheric seeing is not good, it is hard to choose the best focus. The seeing is rarely good for objects near the horizon.

Taking the image

Before starting you need to make sure that the finder is precisely aligned with the main telescope and its CCD camera. This step is very important with CCDs, since their fields are so small: Unless the telescope and the finder are aligned, the chances are that you will not photograph the intended field. Try several exposures. Use different exposure times; for example, the first exposure might be 1 minute in length, the second 2 minutes and third one 4 minutes long. As you display these images, you will undoubtedly find one that is more satisfying than the others. Save this one.

Image calibration

With CCDs, it is almost as important to calibrate your image as it is, when using film, to develop your photographs. Different programs have different ways of accomplishing this calibration. In some, you use a "dark frame," actually a photograph taken with the cover on the telescope, or the camera shutter closed, and lasting the same length as the raw exposure. Others use a "bias frame" which is an exposure of zero seconds. For the best results, it is important to "flat field" each image, a procedure that requires merging the image with a second image of a flat, lit surface (like a white sheet of cardboard or the inside of the dome). Combining the flat-field image with the original should eliminate all the defects in the CCD chip, leaving a clean, evenly textured image. Modern programs like MAXIM DL do this automatically if you have specified images to be used for dark frames and flat fields.

Image manipulation

Co-adding images, a form of image manipulation, is a way of turning a series of short exposures into a single long exposure with a better ratio of signal (real data from the sky) to noise (in the chip). Another way of manipulating an image is "shift differencing," in which two images are superimposed, and then one is rotated a few degrees with respect to the other. A shift-differenced image might reveal faint detail in the coma and tail of a comet. As you gain experience, you will find other imaging tricks that will help increase the beauty and scientific value of your images.

The discovery of Comet Shoemaker–Levy 2

On the night of December 18, 1990 (session 8450AN) Steve Larson and I were photographing comets with our CCD. Included in our list that night were two comets discovered by Jean Mueller at Palomar, and a comet that Gene and Carolyn Shoemaker and I had discovered a few weeks earlier from Palomar Observatory. Also on our list was 1990 UL3, an asteroid that the Shoemakers and I had found a month earlier at Palomar, which follows an orbit that swings out by Jupiter and then back into the inner part of the solar system. Brian Marsden suspected that 1990 UL3 was a comet, but there was no evidence of cometary activity, or fuzziness, in any image of this object. Steve and I planned to take seven 5-minute exposures of the asteroid, as Steve had often done in the past with other newly found asteroids. Next, we would combine all seven images. When the computer presented them as a single "co-added" image, we hoped that the asteroid's motion would give its existence away. The first image appeared as a field full of stars as well as faint lines that showed the weak areas of the chip. Then we flat fielded it, and the improved image contained brighter stars and a darker background; but most important, near the bottom of the field, one of those stars had a tail! The combined images revealed seven positions of the object, which was clearly the one that we had found from Palomar, but the tail, which Steve estimated to be some 70000 kilometers long, clearly revealed its nature as a comet. We reported to the IAU (International Astronomical Union) Central Bureau "that 7 co-added 5-minute exposure Cousins Red band CCD images taken with the Catalina 1.5 m telescope on 1990 Dec 19.3216 mid UT show that the object identified as 1990 UL3 sports a 28 arc sec tail at PA 58 degrees." (The time we sent was for the middle of the 7 exposures, recorded in Universal Time which was 7 hours later, hence the next day.) A few hours later, our message appeared in an *IAU Circular*.[2]

The ability of our CCD system to see deeper than photographs clearly made it possible to determine that this object was a comet. Steve Larson later developed his Catalina Sky Survey, using CCDs in a search for asteroids and comets, and I have developed a CCD system that would work as a comet finder.

Hunting for comets with the Jarnac Comet Survey

Compared with the rigor and effort involved in a photographic comet search, an automated hunt using CCDs seems positively joyful in its ease. We simply create a list of fields to photograph, and the telescope does the rest without the need for any supervision except to open and close the observatory and the telescope. Combined with automated scanning of the images, this literally means that comet hunting can be done in your sleep.

It is not quite that easy, especially at the beginning. Before we could boast of some easy searching, Wendee and I, with a lot of help from programmers and telescope experts, spent almost a year building the system and working out its bugs. Connecting telescope to camera to computer was a difficult process that finally started to work after a lot of trial and error.

Since our current CCD chips do not have very large fields, we have decided to let the sky itself widen the search area. We have cast a net across a portion of sky. Each night we take a strip of 15 photographs for a total area coverage of $\frac{3}{4} \times 7$ degrees. The same area is photographed three times, and we repeat the entire procedure on following nights. Assuming that a typical comet moves half a degree per day, over a month we are essentially covering an area of 15×7 degrees. By photographing two such areas per night, and by using two telescopes, we increase the coverage by a factor of four, to 60×7 degrees. The sky coverage is nowhere near as great as it is with film (we can cover the same area in a single evening with our 8-inch Schmidt camera) but it is more thorough: we can detect comets of 5 magnitudes, or 100 times, fainter. Moreover, since the program mostly runs itself, it has become a part of our daily routine.

Figure 8.1. Using the Spacewatch camera on Kitt Peak in Arizona, Jim Scotti captured the start of the splitting of Comet Takamizawa–Levy (C/1994 G1), discovered in April, 1994, in April, 1995. Courtesy James V. Scotti.

As we put more telescopes and wider-field CCD cameras to work, we increase our area coverage, and this is really the key to discovery in any survey program.

NOTES

1. *Oxford Dictionary of Quotations* (Oxford: Oxford University Press, 1955), 512.
2. *IAU Circular* 5149, December 21, 1990.

9

Comet hunting on paper

... to pore upon a book
To seek the light of truth;

Shakespeare, *Love's Labour's Lost*[1]

Sitting alone in a room in an old building atop Flagstaff's Mars Hill, I was in heaven. It was February 9, 1986, the day that Halley's Comet rounded the Sun, and as I plowed through the old literature, Art Hoag, director of Lowell Observatory, came by to visit. He looked at me, and at the large files of envelopes spread around. "Now that's a new one!" he said, "Comet hunting by reading!"

This chapter adds a twist to the various ways of searching for comets – the concept of searching old photographic plates for comets which might have been missed at the time. It is based on an activity I did with Clyde Tombaugh's Trans-Saturnian planet search plates, and my efforts to report a comet he discovered on them. The story actually began in 1929, when Clyde Tombaugh began his search for trans-Neptunian planets at Lowell Observatory, a program that led to his discovery of Pluto in 1930. For me, it started in 1960, when my father told me the incredible story of how Uranus, Neptune, and Pluto were discovered. Ever since that night I wanted to meet Clyde Tombaugh; when I eventually did, we learned we had much in common in our love of observing the night sky.

In 1985, I began the happy process of interviewing Clyde Tombaugh for a biography I planned to write about him.[2] During this time we reviewed the many aspects of his long and rich life, including the discoveries he had made along the way. As he described for me the variety of objects he had found during his search, he talked of a cluster of galaxies, five open clusters, a globular cluster, many variable stars, and hundreds of asteroids. Then he mentioned a comet. I knew that no Comet Tombaugh was on record. Tombaugh explained that he had found it on plates taken many months earlier, and figuring that the comet was probably faded and nowhere near its discovery position, he decided against reporting it. "Would it be possible to see your discovery images of that comet now?" I asked. The great observer replied helpfully that the detailed notes he had taken on each plate were still preserved at the Lowell Observatory in Flagstaff, Arizona, so all I had to do was to locate the correct plates, and then finding the comet on them would be easy. I asked him when he took the comet plates. His answer was "Sometime between 1929 and 1945." Tombaugh felt the comet was lost months after he took its discovery plates; it was hubris for me to think that I could locate another image of it 55 years later, but at least I could try.

That is what landed me in the plate room in the basement of the main building of Lowell Observatory on February 9, 1986. The room was sparsely furnished with an old wooden chair and a table. One modern addition to the Tombaugh record was an incredible convenience: All the old plate envelopes on which Clyde had recorded his meticulous notes had been replaced with modern archival ones, but the old envelopes were still filed away. It was, therefore, easy to take out several boxes of them, put them on the table, and start searching. On each 14 × 20 brown envelope Tombaugh had neatly listed the numbers of variable stars, asteroids, comets, and other unusual objects detected during his scans. On envelope after envelope I saw his repeated record: "No comets."

I shuddered when I reached the envelope for January 23, 1930, which listed the usual notes, then this at the bottom: "4. No comets. 5. Planet X (Pluto) at last found!!!"[3] I decided to reach for the plate itself and examine history's first discovery plate of Pluto. The second Pluto discovery plate was not in the archive; it rests at the Smithsonian Institution. Then I went back to work, proceeding through the 1930 plates, past the long break after the discovery of Pluto in February, and on into 1931. Late in the afternoon I read the notes on a plate taken in January 1931; after interminable repetitions of "No comets", finally there was one that clearly said the magic word, but phrased as a question: "Comet?"

Could this have been the comet that Tombaugh remembered? I placed the plate under the stereomicroscope and found Tombaugh's marks near the trailed image (it was a 1-hour exposure) of what was doubtlessly a comet. The image was a long, faint, hazy patch of light with a tail attached. Even now only two people knew it existed, Clyde and I, and I was the only one who knew where it was!

I needed some help at this point, so I got a third person involved, observer Brian Skiff on the Lowell staff. We set out to measure the comet's positions on three plates. Then Skiff checked the *Astronomische Nachrichten*, the journal of records for asteroid and comet discoveries of the time, and noted that someone at Lowell Observatory had actually reported it back in 1931, but as an asteroid, not a comet; it had been given the asteroid designation 1931 AN. Whoever reported it – most likely some junior staffer – either had never looked at the image or had no idea of the difference between an asteroid and a comet.

Armed with all this information, we reported the object – correctly as a comet and with accurate positions – to Brian Marsden. He found it interesting, but told us that the object's identity as a comet could not be announced unless an accurate orbit could be determined. For that, I needed more observations than the three we had from January 1931. A few months later, I traveled to the huge collection of photographic plates at the Harvard-Smithsonian Center for Astrophysics; there were plates taken at the correct time, but Comet Tombaugh, it appeared, was too faint to appear on them.

I continued the search over the following 2 years. I visited the great collections at Heidelberg Observatory in Germany, and at Mount Wilson in California. I went to the Meudon Observatory in France, and tried other collections around the world. Heidelberg was the keenest disappointment: Its collection included a plate centered so close to the position of the comet at the time the plate was exposed that I was certain the comet would be there, but it was just off the edge. I encountered the same thing with a plate in the collection at Mount Wilson's Pasadena headquarters. Of course, had the comet appeared on either of these plates, someone would likely have discovered it at the time.

Later that year, Skiff found a second comet marked on a Tombaugh plate; this one later turned out to be a prediscovery image of a comet found at Lowell by Henry Giclas, a colleague of Tombaugh's. There the matter rested for a number of years, but in 1995 I tried a different approach. By writing about and publishing the *process* of this discovery, I reasoned, perhaps some reader might know of the existence of some still-unscanned plate from that era. The article was published in the *International Comet Quarterly*, but it turned up no new plates.[4]

Comet Tombaugh is out there somewhere. For us, it is preserved only as a few photons of light frozen on the photographic emulsion. It is recorded as seven appearances on plates exposed by Tombaugh, three using the A. Lawrence Lowell 13-inch Astrograph, three other images taken simultaneously with a 5-inch diameter "Cogshall" camera mounted atop the big telescope, and a final image at the edge of another Cogshall but too faint to measure accurately. However, there is hope. If the comet is periodic, as appears likely, it might return someday, be discovered again, and finally have its orbit understood. Until then, it's simply C/1931 AN, though in my records it is Comet Tombaugh, a comet discovered by my friend and mentor.

NOTES

1. *Love's Labour's Lost*, 1.1.74–75.
2. See *Clyde Tombaugh: Discoverer of Planet Pluto* (Tucson: University of Arizona Press, 1991).
3. C. Tombaugh, envelope notes for plate No. 171, courtesy Lowell Observatory.
4. D. H. Levy, B. A. Skiff, and C. W. Tombaugh, "Comets discovered by Clyde Tombaugh as part of the trans-Saturnian Planet Search," *International Comet Quarterly*, 17 (1995), 52–53.

10

Hunting for sungrazers over the Internet

But so soon as the all-cheering sun
Should in the farthest east begin to draw
The shady curtains from Aurora's bed,

Shakespeare, Romeo and Juliet[1]

Sunday morning, April 14, 2002, was bright and clear as we began Session *12798S, my daily look at the Sun to count sunspots. Just before I set up the telescope, my friend Tom Glinos and I decided to log on to the Internet to find out what the Sun was up to as seen from space. Thanks to the existence of the Solar and Heliospheric Observatory or SOHO spacecraft, and its coronagraphs, it is now possible to do this. Launched aboard an Atlas–Centaur rocket on December 2, 1995, and currently parked at the L1 libration point between the Sun and Earth, SOHO is about 1.5 million kilometers from Earth in the direction of the Sun. Images are posted every half hour or so on their web site, and that is how Tom and I found ourselves that Sunday morning, peering at the Sun as seen from space. As I excitedly explained to Tom the marvels of seeing an active Sun in a 24-image movie, Tom quietly looked toward the bottom of the screen, pointed out a moving object and said, "What's that?"

Whatever it was, it seemed to be moving in a wide loop toward the Sun, from 6 to 8 o'clock on the screen as the movie of 25 images played over and over. "It's got a faint tail that is getting more obvious as it gets closer to the Sun," I observed. "Tom, you found a comet!" Tom indeed had found a comet, but he was not the first to see it. The first sighting came 2 days earlier through the very same telescope on the same spacecraft, but from an amateur astronomer on the other side of Earth: BoLe city in Xinjiang Province, China. As it entered the field of the LASCO C3 coronagraph telescope, XingMing Zhou found SOHO's 422nd comet. Its official designation is Comet SOHO (C/2002 G3), and it is Zhou's 13th discovery. He uses a 6-inch f/5.3 reflector in these searches.[2]

Zhou comes to the SOHO project with a great deal of experience: "I've hunted the heavens for about 17 years by telescope," he wrote of his work as a visual comet hunter, which has led to two independent comet discoveries.[3] Had his 13th SOHO comet been visible in a dark sky, it would have been as bright as Jupiter.[4]

Comets being found close to the Sun used to be very rare, limited to such anomalies as the eclipse comets of 1882 (Tewfik) or 1948, and the bright comet 1910a found by miners in Africa's Transvaal Premier Diamond mine looking up just out of reach of direct sunlight. When I started comet hunting in 1965, I added a small program designed to look for comets in the direct vicinity of the Sun, and in the "twilight horizon" – quick searches to see if a bright sungrazer was paying a visit. The new approach to finding comets near the Sun began unexpectedly when Comet Howard–Koomen–Michels was found in 1981 in images taken in 1979 from the Solwind spacecraft. Since then more than 500 comets have revealed themselves in this manner, a record that far surpasses every other mode of comet hunting. Almost all of these particular comets are small Kreutz sungrazers: Comets whose orbits cause them to come very close to the Sun, so close that most of them immolate themselves. They are almost never seen when they are far from the Sun. In 2001 Gilbert Esquerdo calculated, for our Jarnac Comet Survey, where the path of Kreutz sungrazers was in the night sky so that we could try to capture comets about a month before perihelion passages. There are only two periods per year when such a search is possible, in the predawn sky in September and October, and in the evening sky in March and April. We used both these observing windows to take long series of CCD images going down to the seventeenth magnitude. We found nothing; even comets that became as bright as first magnitude at perihelion did not appear in our data. These comets must increase exponentially in brightness as they get near the Sun.

In the history of these interesting comets, first discovered by Kreutz in the nineteenth century, three have been found by the Solwind spacecraft, and more through the Solar Maximum Mission (SMM), launched and later repaired by a space shuttle crew; but SOHO has succeeded best – its key being the quality of the three Large Angle and Spectrometric Coronagraphs (abbreviated LASCO) aboard the spacecraft. Invented by the French astronomer B. Lyot in 1930, the coronagraph is a telescope equipped with a device that blocks the brilliant Sun, allowing views of its surroundings; it is especially useful in space where there is no atmosphere to diffuse the Sun's light.[5] Of the three LASCO devices aboard SOHO, C2 and C3 are useful for comet searching, and it is now possible to search for comets using your home computer and the web.

Finding comets the SOHO way

All comets discovered via spacecraft are named for the spacecraft, not the human who discovers them in the spacecraft data, but finding a comet in this way still counts as a comet discovery for the relentless searcher. "You have to be extremely patient and persistent," says Michael Boschat of Halifax, Nova Scotia, who with 32 comets as of the end of 2001 is one of the most persistent searchers. "The main problem now is there are more inexperienced observers looking and more false reports are being posted. And there have been times when even seasoned observers make a false report. Cosmic rays and other defects can move on our images, catching us off guard. You have to take the time to check if your report is real or not."[6] Like anything else worth doing, sungrazer comet searching requires care and patience.

Boschat began his survey by downloading a series of images and then using animation software to loop the images at the rate of four per second, all the while looking "for a point of light moving towards the Sun." It was not long before he discovered his first comet. "I was very excited," he recalled of that day, "and my heart was racing to say the least! The only hard part was noting the x, y coordinates of the comet and getting an email off fast enough before someone else got it."[6]

Profile of a SOHO comet discoverer

Maik Meyer is another amateur astronomer who has had great success finding comets in this manner, and has the additional distinction of having discovered a new group of sungrazer comets. Meyer's interest in discovery grew out of the Jules Verne novels "I read as a kid and – not to forget – Captain Kirk and his crew! Not very scientific but there is the scent of discovery in this TV series." He has observed some 85 comets. "Comets are beautiful, unpredictable, and diverse," he says. "Bright ones are rare and they are connected with the history of mankind like only a very few astronomical events. That alone is motivation enough."[7]

A longtime enthusiast of the Kreutz comet group, SOHO's comet litany was a natural for Meyer. In 1999, at the International Workshop on Cometary Astronomy in Cambridge, England, Douglas Biesecker, who manages the SOHO comet discovery program, encouraged observers to search the SOHO images for comets. "Until the end of 1999," Meyer remembers, "I tried to look occasionally but with no success. I even had problems recovering comets already reported. In 2000, with a fast Internet connection, I searched more intensively.

And then, I had to admit, it became an addiction, one that I have luckily overcome."[7]

Where Meyer stands out is how his interest led to the discovery of a whole new group of comets. The discovery dates to a computer program he wrote that checks his database of comet orbits for similar elements. "My first clear success," he notes, "was the case of comet C/2001 X8 which has the same elements as C/1997 L2." After that, he checked the elements manually and found that a third comet, C/2001 E1, also shared the same orbital elements, except for its inclination to the ecliptic. It turned out that E1's orbit as published was retrograde; Meyer calculated that if the comet was moving prograde instead, the orbit would match the members of his group. The observations of E1 reported do fit the prograde orbit; with virtually no exceptions, the SOHO comets are observed over such short periods of time that their orbits are subject to different interpretation. With three comets confirmed, Meyer set out to find more by testing whether other comets could fit the orbit of his group. He tells his Archimedean story:

> I asked myself if there are not more members. One afternoon I took a bath and suddenly it came to me that the orbits of all previous non-Kreutz SOHO comets can also be wrong. Shortly afterwards I sat at the computer and checked the astrometric data [the precise positions submitted] for the other non-Kreutz comets on similar orbital elements and – Eureka! – three other comets could be represented by the elements in the new group.
>
> I have done a first search for possible progenitors of this group in the annals of ancient comets until now with no success. And I hope for new members – a reason I still search for SOHO comets.[7]

Meyer emphasizes that despite his attraction to SOHO, he will continue to observe comets visually. "All these SOHO comets," he says, "cannot replace the feeling of actually seeing a comet through a telescope or by naked eye."

Searching hints

Finding a comet on a spacecraft image before anyone else is a very difficult process. Unlike other methods of comet hunting, in which most of the sky is available, SOHO comets are in essentially only one field and many people are looking at that one field. A useful web site for searching for these comets is http://soho.nascom.nasa.gov/ – click on the "Latest Images" option under "DATA". Another site is http://sungrazer.nascom.nasa.gov/comets.html, and there is also a link to SOHO on my own web site, www.jarnac.org. According to this sungrazer web site, there are several things to watch out for:

(1) Stars: As the Earth moves around the Sun, the panorama of stars will appear to move past the Sun; if your object is moving in the same direction and at the same rate as the rest of the stars, it is likely a star.

(2) A planet: Planets can move with the stars (from left to right across the image), or against them; bright planets will often leave streaks on the image, an effect of the CCD image called saturation.

(3) A cosmic ray event: This is a high energy particle event, and can appear as a single point, occupying one pixel, or as a streak on the spacecraft image.

(4) SOHO comets: Almost all the SOHO comets follow specific paths that vary from month to month. Brian Marsden has done much work calculating where the sungrazer comets are most likely to be, and the SOHO page has a site that plots these paths, see http://www.ph.u-net.com/comets/c2paths.htm.[8] Your suspect is far more likely to be real if its path is similar to what appears on the appropriate month of this page. However, the bright comet C/2002 G3, described at the opening of this chapter, was not a Kreutz sungrazer and did not follow a typical sungrazer route.

On searching for sungrazers, Maik Meyer says:

I still use only Netscape as a tool to blink three or four consecutive images. The main thing is to be fast; chances are low to find a comet in images as old as a day. Newcomers often report comets in images taken 1 week or so ago, and given the close surveillance by the experienced hunters, such a find is unlikely to be real. Non-Kreutz comets are harder to find, especially when faint.[7]

When he finds a comet – and he has found 26 to the date of this writing – Meyer notes the *x, y* coordinates in pixels and quickly posts his claim on the comet report page. Despite all the hints and suggestions that are available, comet hunter Sebastian F. Hönig insists that "there's one thing you have to get yourself: Experience! That's perhaps the most important tool…. Just search for every comet – faint or bright – which is reported."[9] I also recommend his excellent training web site at http://www.sungrazer.org.

Despite the care you take, this type of comet hunting is difficult since so many others are searching as well. It seems easy on the surface – comet hunting over the web from the comfort of your living room – but it has its challenges. However, I think it is absolutely amazing that we have this option. If someone had told me about this brave new world of comet hunting back in 1965, I would not have believed it. SOHO, its instruments, and the Internet, have provided yet another way to make observational astronomy, and the possibility of discovery, varied and interesting.

SWAN comets

SOHO also carries an instrument called SWAN (for Solar Wind ANisotropies). It images much of the sky in the wavelength of Lyman alpha, In July, 2002, M. Suzuki discovered a comet on these images. If a comet is brighter than magnitude 10 (see Chapter 16 for a discussion of magnitude) and has sufficient gaseous emissions, it can be discovered in this manner.

NOTES

1. *Romeo and Juliet*, 1.1.134–136.
2. *Spaceflight Now*, posted April 19, 2002. http://spaceflightnow.com/news/n0204/ 19sohocomet/.
3. X.M. Zhou to D. Levy, June 11, 2002.
4. The original discovery information on the SOHO reports page reads as follows:

> Apr 12 2002 06:37:38
> possible non-group comet in real-time c3 images now.
> with short tail, from 04:18–09:42
>
> 0,0 in upper left corner.

Images	[Size]	[X, Y]
20020412.0418.c3.gif	512 × 512	215, 507
20020412.0542.c3.gif	512 × 512	214, 506
20020412.0642.c3.gif	512 × 512	213, 505
20020412.0742.c3.gif	512 × 512	212, 504
20020412.0842.c3.gif	512 × 512	211, 503
20020412.0942.c3.gif	512 × 512	209, 503

> XingMing Zhou

From the *SOHO–LASCO Comet Report Archive, SOHO–LASCO Comet Reports for 200204* at http://sungrazer.nascom.nasa.gov/comets_found.old/comets_2002/ comets200204_arch.html.
5. See B. Lyot, "La Couronne solair étudiee en dehors des eclipses" *Comptes Rendus Helodomadaires des Séances de L'Académie des Sciences* (Paris) 191 (1930), 834.
6. M. Boschat to D. Levy, May, 2002.
7. M. Meyer to D. Levy, June 5, 2002.
8. Brian Marsden, *Astronomical Journal*, 98:6 (1989), 2306–2321.
9. *Sebastian's Comet Hunt*, http://www.sungrazer.org/.

Plate I. Comet Levy 1990c (C/1990 K1) shown through Terence Dickinson's
4-inch reflector in his Ontario observatory, in August, 1990. Photograph courtesy
Terence Dickinson.

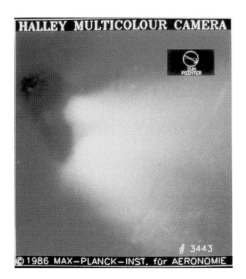

Plate II. The nucleus of Halley's Comet (1P/Halley) as seen through the Halley Multicolour Camera, Giotto Spacecraft, March, 1986. Courtesy European Space Agency at Max-Planck-Institut für Astronomie (H.U. Keller).

Plate III. Comet Levy (1990c (C/1990 K1), shown in this 22-minute exposure through Terence Dickinson's 4-inch refractor in August, 1990. Photograph by Terence Dickinson.

Plate IV. Comet Ikeya–Zhang (C/2002 C1) photographed by Scott Tucker of Starizona, on March 12, 2002, near Vail, Arizona, using an Astro-Physics 155EDF telescope and Pentax 67 camera. Courtesy of Scott Tucker.

Plate V. Comet Snyder–Murakami (C/2002 E2) photographed by Tim Hunter, through a 12-inch Meade Schmidt-Cassegrain + CCD, at his 3towers Observatory in Arizona, in March, 2002. Courtsey Tim Hunter.

Plate VI. Comet Hale–Bopp (C/1995 O1) photographed by Dean Koenig of
Starizona, Tucson, Arizona, in March, 1997. Courtesy Dean Koenig.

Plate VII. Comet Hale–Bopp (C/1995 O1) photographed in March, 1997 at his 3towers Observatory in Arizona by Tim Hunter, using a 17-mm f/4.5 lens (the red lights are on the towers, which form part of a regional 91 (FM) radio system). Courtesy Tim Hunter.

Plate VIII. Comet Hyakutake (C/996 B2) in a 90-minute exposure, taken by the author through a Yashika twin-lens camera, in Arizona, in March, 1996.

11

What to do when you think you have found a comet

Then felt I like some watcher of the skies
When a new planet swims into his ken;

John Keats, *On First Looking into Chapman's Homer*[1]

Much as the search for comets is exciting in itself, the ultimate goal of the search, of course, is to discover a comet. The chances are that you will find one – or at least you will think you have found one – after just a few days. In all likelihood, you will have found one of the sky's many "comet masqueraders" – as Leslie Peltier called them – a galaxy, nebula, or cluster that resembles a comet.[2]

The larger your telescope, the more comet suspects you will find and the more time you will need to spend checking out suspicious objects. You can locate the object you see on some atlas, like Wil Tirion and Roger Sinnott's *Sky Atlas 2000.0*, or Tirion's *Uranometria 2000.0*, or *Sky & Telescope's Millennium Star Atlas*.[3]

An encoder system on your telescope, like a "goto" mount or the NGC-MAX or Meade's Magellan available for altazimuth telescopes, is a much easier way to keep track of where you are. I have learned not to trust these devices wholeheartedly, however: mine fooled me once, saying that no object was present at the indicated position when, indeed, the position was simply off. Also, if the object being observed has no listed magnitude, it will not show up unless you set the magnitude setting to the maximum.

Discovering a comet

If your atlas or goto device shows no fuzzy object where there is clearly one in the field of your telescope, then follow the following steps:

(1) Try to determine the nature of the suspect. Could it be a ghost image of a nearby bright star? Try moving your telescope a bit. If you have a ghost

image, it should change its position relative to nearby stars as you move the telescope. Could it be a faint star or group of faint stars? Use high power to rule this possibility out. Does the object have a tail? Although most faint comets do not show obvious tails, there are very few objects in the sky that are not comets that do have tails; Hubble's Variable Nebula is an example.

(2) Draw a sketch of the field of view centered on your suspect.

(3) Draw a second sketch of the field of view of the finder or other sighting device as insurance against losing it.

(4) Wait to see if the object is moving. I continue comet hunting during this time, going back in 15 minutes or half an hour to check if the suspect has moved.

(5) If there has been no motion by dawn of the setting of the suspicious object, I advise strongly against informing the Central Bureau for Astronomical Telegrams. All comets do show motion eventually. You should wait to confirm motion, even if that process means you must wait another 24 hours. For the comet I found in 1988, I actually observed it over three mornings before being confident enough to send a comet message. If the object has moved, stay calm! The next minutes will be busy and critical ones.

(6) Sketch the object again and note its position.

(7) Could your suspect be a known comet? Check even the positions of comets that should be too faint for your telescope, although if you should be the first to notice an unexpected increase in a known comet's brightness, you definitely should report it. The positions of the brighter comets are published in the major astronomy magazines, while the *IAU Circulars* published by the International Astronomical Union, the *Minor Planet Circulars* published by the IAU Minor Planet Center, and the annual *Comet Handbook* published by the *International Comet Quarterly* are good sources to check for known comets. The web site Minor Planet Checker at http://scully.harvard.edu/~cgi/CheckMP is also an excellent and up-to-date source, or, if you subscribe to the computer service of the Central Bureau for Astronomical Telegrams, you can ask their program to identify known comets or asteroids in the area around your suspect.

(8) If possible, have your sighting confirmed by an experienced comet observer. Tell that person the position, suspected nature, and direction of motion of the object.

(9) Once you do detect motion, have ruled out previously known objects in the area, and have had your observation confirmed locally, it is time to notify the Central Bureau for Astronomical Telegrams (CBAT) in

Cambridge, Massachusetts. Its email address is: cbat@cfa.harvard.edu, or, you can use the web site at http://cfa-www.harvard.edu/iau/cbat.html.

Send a follow-up message by mail or courier to the Central Bureau for Astronomical Telegrams, 60 Garden Street, Cambridge, MA, 02138. Its director, Daniel Green, needs the following information:

A. Suspected nature of object (e.g., a comet).
B. Right ascension and declination in 2000.0 coordinates.
C. Direction and rate of motion. Preferably, this is supplied with a second position taken a half hour or more after the first. This motion is *very* important, since it may be some time before a clear sky somewhere permits another observer to confirm your finding.
D. The comet's overall magnitude (see Chapter 16).
E. Whether the discovery was a visual or a photographic one. If you observed the object both ways, clearly distinguish how you made each observation.
F. Your description of the object's appearance, including remarks about its size, angular diameter, and shape and length of the tail if you see one.
G. The date and time of your observation, converted to Universal Time, or at least with the time zone included.
H. Details of the instrument you used. Include the telescope's aperture, type, and magnification. If you are reporting a photographic find, add the film emulsion used, or for electronic observations, the CCD camera model; also add the exposure time, the limiting stellar magnitude of the photograph, and the size of the field in degrees or minutes of arc.
I. Your name as discoverer, your postal address, email address, and telephone number.

N.B.: *Do not send a message unless you are certain that the object is a new comet.* Imagine the chagrin of the observer I know when he reported having "discovered" the Andromeda galaxy. According to the director emeritus of the CBAT, Brian Marsden, approximately 98% of comet discovery reports from unknown observers turn out to be false alarms.[4] Just because an object is fuzzy does not mean that it is a comet, and even if it is a comet it could be a known one. Galaxies, nebulae, and ghost images of bright stars are often reported as new comets. Also, *never* report a fuzzy spot on a single photograph without some other observer's confirmation.

How comets are designated and named

The system for how comets are designated was changed a few years ago. Under the old system, a comet was designated according to its order of discovery or recovery in a given year (e.g., 1982i, Comet Halley, was the ninth comet to appear in 1982, and Comet Levy 1987y was the 25th comet to be reported in 1987). After cometary information was complete for a given year, each comet was also assigned a Roman numeral designation indicating the numerical order of its perihelion passage (e.g., Halley at its last return was also known as 1986 III).

At the International Astronomical Union's 22nd General Assembly in The Hague in 1994, Commission 20 approved a resolution that changes the way comets are designated. Under this new system, which came into effect in January 1995 but which is applied retroactively to every comet for which a reasonable orbit is available, a comet is assigned only one designation. To accomplish this, the year is divided into periods called "half-months" beginning with A (the first half of January), omitting the letter I, and concluding with Y (the last half of December). In addition, there is a letter indicating the comet's status: C for a long period comet (defined as having a period of 200 years or longer), P for a "periodic comet" (defined as having a period of less than 200 years), X for a comet for which a reasonable orbit cannot be computed, or D for a disappeared or deceased comet. The half-month letter is followed by a consecutive numeral to indicate the order of discovery announcement during that half-month. In addition, once the orbit of a periodic comet is well known, typically after it has been observed on more than one return, that comet receives a permanent number according to the order in which the comet's periodicity was recognized. Thus, long-period Comet Hale–Bopp, as the first comet to be found in the "O" part of 1995, is labeled C/1995 O1. Deceased Comet Shoemaker–Levy 9 was originally 1993e; its new designation, as the second comet to be found in the "F" part of 1993, is D/1993 F2. The new designation for periodic Comet Halley is 1P/Halley, and that for periodic Comet Shoemaker–Levy 2, 1990p, is 137P/Shoemaker–Levy 2.

When a comet becomes well known, the vast majority of scientists, the press, and public ignore the cumbersome new designation, preferring instead to use the more easily remembered proper names. Our experience with comets Hale–Bopp and C/1996 B2 (Hyakutake) showed clearly that people were far more comfortable with their proper names alone. However, with more than 200 comets named SOHO, and more than 50 named LINEAR as this chapter is written, we need to become familiar with the designation in order to separate one comet from the others.

A handful of comets are named not for their discoverers but for those who computed their orbits. The most famous of these is Halley's Comet. Two others are 27P/Crommelin, and 2P/Encke. Encke's comet was named to honor the German mathematician Johann Franz Encke's work in uncovering the rapid motion of this interesting comet. The comet's story began on January 17, 1786, when the French comet hunter Pierre Méchain found a comet of fifth magnitude, barely visible to the naked eye. Nine years later, on October 20, 1805, the English comet hunter, Caroline Herschel, discovered the comet, now magnitude 5.5, as it returned. Also on October 20, 1805, another French comet hunter, Jean-Louis Pons, discovered it for a third time, again at magnitude 5.5. By this time comet hunting was quite a popular activity, and two other observers, Johann Huth and Alexis Bouvard, independently found the comet shortly after Pons. At the 1805 passage, Encke calculated an orbit for the comet, and he was surprised to suggest that it might be a comet that returns every 12 years. On November 26, 1818, Pons once again discovered this comet. Encke then connected Pons's new comet to the 1805 comet, but in so doing he realized that its period was not 12 years, but $3\frac{1}{3}$. Encke's calculations showed that Pons's twice-found comet was the same as those of Méchain and Herschel. On June 2, 1822, the Australian searcher Charles Rumker recovered the comet, confirming Encke's work once and for all. Since then it has been followed on most of its returns. On the morning of August 9, 2000, I picked it up while comet hunting. Although the comet is now well-known and famous, it was nevertheless a thrill to greet it as it returned for one of its many visits.

If you should have the rare fortune to discover a comet, you really are following in the footsteps of Messier, whose many comet finds launched the tradition of naming comets for their discoverers. You can also share the feeling of Keats, two centuries ago, as he felt the thrill of a watcher of the skies.

NOTES

1. H. W. Garrod (ed.), *Keats: Poetical Works* (London: Oxford University Press, Second Edition, 1966), 38.
2. Leslie C. Peltier, *Starlight Nights: The Adventures of a Star-Gazer* (New York: Harper & Row, 1965), 228.
3. Wil Tirion and Roger W. Sinnott, *Sky Atlas 2000.0* (Cambridge: Cambridge University Press, Second Edition, 1998); Wil Tirion, ed., *Uranometria 2000.0 Volumes I and II* (Richmond, VA: Willmann-Bell, 2001); Roger W. Sinnott and Michael A. C. Perryman, *Millennium Star Atlas* (Cambridge, MA: Sky Publishing, 1997).
4. B. Marsden, personal communication, July 17, 1993.

PART III

A new way of looking at comets

12

When comets hit planets

One of my favorite memories took place during Stellafane
PreConvention, in 1994. It was near dusk on Sunday, July 31st. Tom
Spirock and I went up to the Porter Turret telescope to find John just
finishing giving a tour of Stellafane to some of the locals. As the sky was
darkening, John guided the turret towards Jupiter. All of a sudden a
string of expletives I won't repeat flowed from his mouth! I thought the
primary mirror had shattered. Then he just stared . . . I sat there waiting,
then when John moved away from the eyepiece, I saw what had set him
off. There in the eyepiece, floated the scarred face of Jupiter, a black band
starting to form from the impacts of Comet Shoemaker–Levy 9. The
Hubble views were amazing, Calar Alto was exciting, even the views
from my backyard were memorable, but seeing this, through a quality
instrument such as the Porter Turret, will always be with me. As I stared,
John began recalling the details, I confirmed each one as he said it. For
the next 20 or 30 minutes the three of us kept taking turns at the
eyepiece, soaking up the views that we knew we'd probably never see
again. John was nearly moved to tears.

Wayne Zuhl, Springfield Telescope Makers, Stellafane Convention and Observatory,

Vermont, 2002

In a cosmic sense, the collision of the ninth periodic comet discovered
by the team of Carolyn and Gene Shoemaker and David Levy with
the planet Jupiter was unremarkable. The history of the solar system,
indeed its very genesis, has been marked by countless such events. . . . In
human terms, on the other hand, the impact of Comet Shoemaker–Levy
9's 20-odd pieces with Jupiter was an unprecedented event of global
significance. . . . For a week in July, the world looked up from its normal
preoccupations long enough to notice, and to ponder, the awesome beauty
of the natural world and the surprising unpredictability of the universe.

Keith S. Noll, Harold A. Weaver, and Paul D. Feldman, 1996[1]

> . . . Stars, hide your fires;
> Let not light see my black and deep desires.

Shakespeare, Macbeth[2]

The impact of Comet Shoemaker–Levy 9 with Jupiter

As building blocks for larger worlds throughout the galaxy, comets are very important. When a big comet passes by, telescopes on Earth and in space are pointed toward it to try to determine its structure, composition, and course. In the history of astronomy, no single event was studied with more energy and telescopic power than the collision of a small comet called Shoemaker–Levy 9 with Jupiter. The event taught us all about what happens when a comet becomes a projectile, and collides with a planet.

Although each comet is unique, Shoemaker–Levy 9 provided surprises from the start. On the evening of March 23, 1993, Gene and Carolyn Shoemaker and I took two photographs of a region of Jupiter as part of a comet and asteroid search program being conducted at Palomar Mountain in southern California. By the afternoon of March 25, having completed examining the films we had taken earlier, Carolyn placed the two films encompassing Jupiter in her stereo-microscope, the instrument we use to scan pairs of films for new comets and asteroids. As she scanned down and across the films, Carolyn passed over the brilliant image of Jupiter itself, and she continued on.

It is interesting how major discoveries often begin not with drums and fireworks, but with a strange look or expression. Gene and I had just returned from a drive through the wind and blowing snow to collect the night's film. We returned to the dome of our 18-inch telescope quite pessimistic that we would have any observing during the coming night. Gene sat down to read, and I returned to working on some writing, while Carolyn continued to scan our Jupiter film. She actually passed over a strange elongated fuzziness, and it was almost out of the field of view of her stereomicroscope before she quickly moved the object back into the center of the field. She sat up tensely. "I don't know what this is," she declared. "It looks like a squashed comet."

Indeed, it did look like a comet, but instead of the asymmetrical coma that comets should have, this one displayed a bar of cometary light. What is more, in the hour and 49 minutes between films the whole thing had moved only slightly. "The image is most unusual," we reported to Brian Marsden, then director of the International Astronomical Union's Central Bureau for Astronomical Telegrams, "in that it appears as a dense, linear bar very close to 1 arc-minute long, oriented roughly east–west. No central condensation [a starry point sometimes marking the comet's center] is observable in either of the two

1993 e 1993 03 24.43072 1993 03 24.35503 N

Figure 12.1. The two discovery images of comet Shoemaker–Levy 9 (D/1993 F2) photographed by David Levy, Gene and Carolyn Shoemaker on the evening of March 23, 1993, through the 18-inch Schmidt camera at Palomar Mountain in California.

Figure 12.2. Comet Shoemaker–Levy 9 (D/1993 F2) photographed by Jim Scotti
through the Spacewatch telescope on Kitt Peak in Arizona in March, 1993.
Courtesy James V. Scotti.

images. A fainter, wispy "tail" extends north of the bar and to the west."[3] Finally, each side of the bar ended with a long, thin line.

Because the sky was so cloudy, we could not get a confirmation image ourselves. Instead I telephoned Jim Scotti, who was that very night observing at
the Spacewatch camera atop Kitt Peak, some 400 miles east of us. As I expected,
the storm that blinded us to the sky had not reached him, and although he was
working on his own program he did promise to try to confirm our discovery.
Two hours later, I telephoned him again. "Do we have a comet?" I asked. "The
sound you just heard," Jim answered excitedly, "is me trying to lift my jaw off
the floor. I can see at least five separate cometary condensations, but there are
probably many more. And on either end is a long dust trail. Boy, do you three
ever have a comet!" At this point we knew we had found a real unicorn in the
astronomical zoo, for never before had any comet been observed that had been
so catastrophically disrupted. Soon we learned the reason for the disruption, a
close tidal interaction with Jupiter, the solar system's largest planet. The comet,
it turned out, had grazed by Jupiter just a few months before discovery. Just as
the Moon pulls on the waters of Earth, Jupiter's gravity stretched this comet so

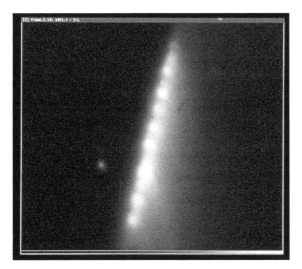

Figure 12.3. Comet Shoemaker–Levy 9 (D/1993 F2) photographed through the 90-inch telescope at Steward Observatory in Arizona, by Wieslaw Wisniewski, in March, 1993. Courtesy James V. Scotti.

strongly that it fell apart into many fragments, a "string of pearls" each with its own tail (see Figure 1.2)!

The comet's strange appearance was only the herald of what was to come. Instead of orbiting the Sun, as every other known comet has done, Shoemaker–Levy 9 was tracing a path around Jupiter, and had been since about 1929. On May 22, 1993, the International Astronomical Union issued its now-famous announcement that Comet Shoemaker–Levy 9 was in its final orbit, and would collide with Jupiter during July, 1994.[4] That gave the astronomical community 14 months to plan the biggest observational campaign ever put together in the history of astronomy for a single event.

By the night of the first impact on July 16, 1994 (exactly 7 years before I wrote these words), virtually every major telescope on Earth was pointed towards Jupiter. In space, the Galileo spacecraft en route to Jupiter, and the Hubble Space Telescope, watched as a fragment of Shoemaker–Levy 9, traveling at 134000 miles per hour, tore into Jupiter's atmosphere. By the end of "crash week," Jupiter's southern hemisphere was blackened by dark clouds that would persist for months. In this particular war of the worlds, the tiny comet obviously lost, but Jupiter was injured.

What did the impact teach us?

More important than any single bit of information was this: The impact of Comet Shoemaker–Levy 9 with Jupiter reminded us that comets strike

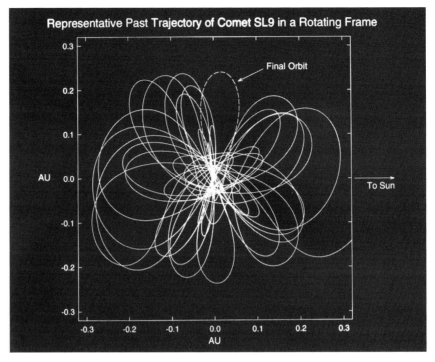

Figure 12.4. An intriguing history: One of the possible orbit diagrams showing how Comet Shoemaker–Levy 9 (D/1993 F2) winged its way around Jupiter. Courtesy Donald K. Yeomans and Paul Chodas, NASA/JPL.

planets, a kind of event that humanity had never seen before. In the early years of our solar system, comet-like bodies called planetesimals hit each other to create and build up our system of planets. After the planets were formed, comets continued to pound them, bringing with them their precious organic materials and, after a long process, eventually setting off the origin of life on at least one planet. Other comets changed the course of life through the heroic forces of their collisions: 65 million years ago, a collision with a comet, or possibly an asteroid, ended the era of the dinosaurs, opening the way for mammals and humanity. The breakup and collision of tiny comet Shoemaker–Levy 9 showed us how this process worked.

How rare is a comet–Jupiter collision? The answer depends in part on how large Comet Shoemaker–Levy 9 was before it broke apart in 1993. The comet could have been as small as a kilometer, or as large as 10 kilometers, in diameter. If the comet was near the lower end of that range, say about 1.5 kilometers in diameter, then according to discoverer Gene Shoemaker we would see a single collision every century or so. Such a comet hitting Jupiter should have left a massive dark spot, probably larger than any of the

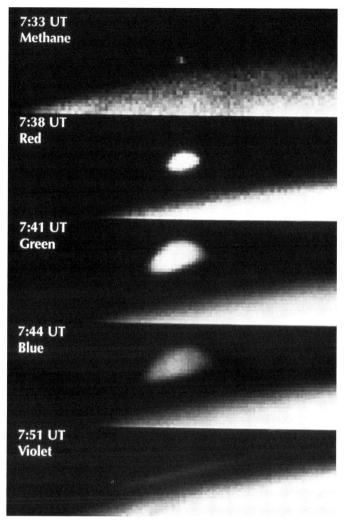

Figure 12.5. The changing plume over Jupiter resulting from the impact of Fragment G of Comet Shoemaker–Levy 9 (D/1993 F2), July 18, 1994, taken through the NASA/Hubble Space Telescope. NASA/HST.

Shoemaker–Levy 9 spots; but statistics are figured over the long term, and Shoemaker–Levy 9 could well have been the first major impact in a long time. Half of these colliding comets should be in an orbit around Jupiter when they collide with the planet so, statistically, such a crash should take place every 200 years. If the comet broke apart, and later collided with Jupiter, as Shoemaker–Levy 9 did, then collision events would be far less frequent. If the breakup happened two or more orbits before impact, the individual pieces would long since have gone off on separate orbits, with some hitting Jupiter at widely separated intervals, and others escaping. The most

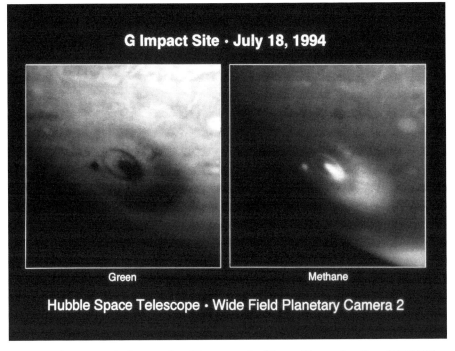

Figure 12.6. The impact site of Fragment G of Comet Shoemaker–Levy 9 (D/1993 F2) on Jupiter July 18, 1994, taken through the NASA/Hubble Space Telescope. NASA/HST.

optimistic frequency for the multiple-impact event we saw in 1994 is once every *2000 years*.

The impact taught us many other things. One was the appearance of dark spots over each impact site, an effect that no one had predicted. Some investigators concluded that the dark spots were comprised of hydrocarbons (of which soot is a familiar form). Others thought they were made up of silicaceous material. The spots could be used as the signature effect of impact: If we go back through past observations of Jupiter, would we find other examples of this? According to Thomas Hockey, who researched historic observations of Jupiter for his Ph.D. dissertation at New Mexico State University, there have been other instances of dark spots, but none as prominent as these.[5] Hockey did uncover a few examples of spots that date back to 1690, when the famous astronomer Giovanni Cassini observed a large spot in Jupiter's south equatorial belt. Isshi Tabe, a Japanese amateur astronomer who studied historical records of Jovian atmospheric phenomena, found a series of drawings by Cassini in the archives of the Paris Observatory that seemed to anticipate the more moderate Shoemaker–Levy 9 impacts (like those of fragments A, C, E, and H). Cassini followed the spot for 18 days, seeing it early in December 1690, and following it until December 23.[6] Almost a century later, from his home in England in 1778, William Herschel drew three large equatorial spots on Jupiter, each of which occupied less than a tenth of the planet's diameter. Half a century after that, in 1834, George Airy, also from England, described "a remarkable spot seen on the apparent southern belt, nearly four times as large as the shadow of the first satellite."[5] In 1850, William Dawes and William Lassell recorded a series of spots in the South Temperate Zone, "two being nearly equal in size, and almost as large as the third satellite appears."[5] Amateur astronomer Steve O'Meara has also studied old visual observations of the planets for many years. He cautions that it is difficult to compare drawings of dark spots with modern photographs, since drawing techniques often resulted in any dark spots appearing darker than they really were.[7] The Shoemaker–Levy 9 impact spots, Hockey concluded, "would have amazed the great planetary observers of the 17th, 18th, and 19th centuries."[5]

The comet crash attracted wide general interest. Both *Time* and the *U.S. News & World Report* did cover stories about the impact and its consequences, while planetary scientists continued to study it for years afterward. One of the most unexpected effects was that Jupiter showed evidence of the impacts for some time after the third week of July. The dark spots lasted for several months, spreading out into a large dark belt surrounding the southern hemisphere impact latitude of the planet. As late as 2001 – more than *7 years* after the impacts, carbon monoxide was not only still being detected in Jupiter's atmosphere, but it was also spreading past the equator into the planet's northern hemisphere.[8] Clearly, the story of the unfortunate comet Shoemaker–Levy 9 and its effects on the solar system's biggest planet goes on.

NOTES

1. Keith S. Noll, Harold A. Weaver, and Paul D. Feldman (eds.), *The Collision of Comet Shoemaker–Levy 9 and Jupiter* (Cambridge: Cambridge University Press, 1996), xiii.
2. *Macbeth*, 1.4.50–55.
3. E.M. and C.S. Shoemaker and D. Levy to Brian Marsden, March 25, 1993.
4. *IAU Circular* Nos. 5800 and 5801, May 22, 1994.
5. Thomas Hockey, University of Northern Iowa, email to the SL9 Message Center, University of Maryland, July 22, 1994. See also "A Historical Interpretation of the Study of the Visible Cloud Morphology on the Planet Jupiter: 1610–1878" (Doctoral Dissertation, New Mexico State University, 1988).
6. I. Tabe, J. Watanabe, and M. Jimbo "Discovery of a possible impact spot on Jupiter recorded in 1690," *Publications of the Astronomical Society of Japan* (Letter), 49:1 (1997).
7. S. O'Meara, *Sky & Telescope*, personal communication, August 15, 1994.
8. B. Bézard, E. Lellouch, D. Strobel, J. P. Maillard, P. Drossart, "Carbon monoxide on Jupiter: Evidence for both internal and external sources," *Icarus*, 159 (September 2002), 95–111.

13

The future of visual comet hunting

In spite of this increasing competition there always will be comets for the amateur to seek and, in some facets of this work, he still has an advantage. In a given time he can cover far more sky than can the camera, he can know within half an hour the true nature of a suspected object and he can search much closer to the sun in regions which would fog a photographic plate.

Leslie Peltier, *Starlight Nights*, 1965[1]

Over time, amateur astronomers who use wide-field CCDs will be able to continue discovering comets; but what about amateur astronomers who do visual searching? Comet hunting has utterly changed in the 37 years since I started doing it in 1965. When I found Comet Takamizawa–Levy in 1994, it still seemed possible to conduct a visual comet search in the tradition of eighteenth-century French master Charles Messier, his British counterpart Caroline Herschel, or Leslie Peltier. It was the efforts of skywatchers like these that inspired me to begin comet hunting that year.

Back then, the threat to visual searches was from photographic surveys, which were succeeding in finding most of the comets. However, these efforts were concentrating on the sky at opposition to the Sun, and not terribly thoroughly at that, leaving many comets available for visual observers. Even after the enormous productivity of the great Palomar photographic surveys of the 1980s and 1990s – surveys specifically designed to discover comets and asteroids – until a few years ago it still seemed possible for comets to be found visually.

In the year before I completed this chapter (in May 2002) the International Astronomical Union's Minor Planet Center recorded at least 125 comet appearances. Almost three-quarters of those comets have the same name SOHO, and as we have already seen, these are comets discovered using the Solar and Heliospheric Observatory. These comets are "extra" – objects that could not have been discovered by visual observers, or for that matter, by any other process known at this time. There were 17 comets found by a U.S. Air Force and NASA-funded project named LINEAR (Lincoln Near Earth Asteroid Research), and a handful of comets discovered by other programs. Possibly two or three of these comets, but not many, could have been found by visual observers months later if the professional surveys had missed them.

What about old-fashioned visual discoveries during this time? Comet Petriew, discovered by Canadian amateur astronomer Vance Petriew during a star party in August 2001, was the first visual find in some time. Then on February 1, 2002, Comet Ikeya–Zhang (see Plate IV) turned up as part of the visual comet searches of Kaoru Ikeya and Daqing Zhang; Brazilian Paulo Raymundo also found it visually the same night. Brighter than eighth magnitude, this comet was an invitation to the multiple discovery it received.

Six weeks later, on March 11, Douglas Snyder, from the Sierra Vista area of southeastern Arizona, and Shigeki Murakami from Niigata, Japan, discovered a comet traversing the Milky Way high in the eastern predawn sky (see Plate V). Only a week after that, on March 18, Syogo Utsunomiya from Kumamoto-ken, Japan, discovered a comet also in the eastern predawn sky. Does this mean that the lack of amateur-discovered comets in the late 1990s and early 2000s was a statistical low? According to Daniel Green of the Central Bureau for Astronomical Telegrams, it is the three comets in early 2002 that were the anomaly. "We are anticipating the time in the near future when amateur astronomers are not going to be discovering comets any more," Dan said recently. "At some point there will be no more amateur discoveries."[2] If he is right, the numbers of visual discoveries will continue to decline almost to zero.

Figure 13.1. Comet Utsunomiya (C/2002 F1) photographed by Tim Hunter, through a 12-inch Meade Schmidt-Cassegrain + CCD, at his 3towers Observatory in Arizona, in March, 2002. Courtesy Tim Hunter.

Because the major surveys do not cover the entire sky, especially the area near the Sun, it is still possible to find a comet visually, and some comets happen to come in at a shallow angle to the Sun that hides them from the big surveys; but the date and time of the last visual comet discovery is approaching. When will that discovery take place? I do not know. We have already noted (see Chapter 7) Peltier's 1966 letter to me in which he asserted that most finds were not visual. Peltier might have predicted back then that visual discoveries would vanish, but as this chapter's opening quote attests, he did not. Indeed, 18 years after I received his letter, I found my first comet, and have found seven more since.

In this chapter's message to new and experienced comet hunters, I will take Peltier's side. I still believe that there will always be comets for the amateur to seek and find through his or her visual telescope. It is harder than it used to be, and many searchers will give up. In 1967, Robert Burnham Jr., who discovered comets both visually and photographically, advised me that "If you hunt long enough, stay away from the galaxies in Virgo, and never give up, some day you will find a comet."[3] I think this advice still holds, although that "some day" could be decades into the future for most searchers.

As you hunt for comets, remember that it is not a good idea to have the discovery of a new comet as your only goal. In 1997, Leif Robinson, then the editor of *Sky & Telescope*, came down pretty hard on single-minded searchers. "I've never had any great admiration for comet hunters," he editorialized, "To spend hundreds of hours in failure for each minute of success never seemed like a good deal to me.... For amateurs there's the allure of getting your name hitched to a star, albeit a hairy one. If you're *very* lucky, like Thomas Bopp, your name might appear in textbooks for years."[4] If the only reason you spend all this time comet hunting is to find a comet, then Robinson has a point. So did the great Japanese comet hunter Minoru Honda, who discovered 12 comets and 11 novae during his lifetime, in the advice he gave Kaoru Ikeya before the young comet hunter made his first discovery: "If you desperately want to find a new comet, please stop your search because you may never be able to find a new comet. However, if you are content to search the sky without ever experiencing a new comet discovery, please keep searching because someday, you may be able to find a new one."[5]

There are two kinds of visual observing, the "star party mode" and the "comet search". In the star party mode you stand in line (maybe a line of one, but still a line) and look through a telescope at object after object. You decide, or at least the owner of the telescope decides, what object you will next look at. Comet hunting, however, is a different way of observing the night sky. At the start of each session, I look over the area of sky through which I plan to search. With little idea of what to expect in the next hour or so, I like to imagine

Figure 13.2. Comet Ikeya-Zhang (C/2002 C1) photographed by Tim Hunter, through a 12-inch Meade Schmidt-Cassegrain + CCD, at his 3towers Observatory in Arizona, in March, 2002. Courtesy Tim Hunter.

a celestial stage opening up for me. The objects I will see are the sky's choice, not my choice. "Go ahead," I say to the sky, "Make my night." With that approach, every hour spent comet hunting is a time for discovery.

NOTES

1. Leslie C. Peltier, *Starlight Nights: The Adventures of a Star-Gazer* (New York: Harper & Row, 1965), 230.
2. *Let's Talk Stars* radio show, KTKT Tucson, Arizona, April 30, 2002.
3. Robert Burnham to David Levy, June 6, 1967.
4. Leif Robinson, "Ticket to nowhere?" *Sky & Telescope*, 94 (August 1997), 8.
5. D. Levy, "The Comet Master." *Sky & Telescope*, 104 (July 2002), 70. Original quotation translated by Shigeru Hayashi.

PART IV

How to observe comets

14

An introduction to comet origins and characteristics

I have believed that the comet, shining on all directly from the same place and appearing the same from all sides, must be considered as worthy of the heavens and very near to the stars.

Horatio Grassi[1]

Observing comets is unlike observing anything else in the sky, because comets are unlike anything else in the sky. They are in our solar system, yet they resemble in appearance the galaxies that shine from vast distances. They move undisturbed through the sky, as meteors do, yet their motion is much slower and their appearance is utterly unlike that of a meteor. They change in brightness like variable stars, yet they are not stars.

In a sky filled with strange and ill-behaved objects, comets hold their own as worthy of interest, study, and observation, but in the centuries that comets have been observed, surprisingly little has been learned about the details of their origin and evolution, and of the processes occurring in them. Most of our detailed knowledge of comets has been acquired over the past few decades, and a good proportion of that in the last few years.

How were comets formed?

When the solar system was forming, so the most favored theory goes, comets were formed in the region moving outward from Uranus. The evidence for this idea lies in the observation, made during Comet Halley's visit in 1986, that its ratio of deuterium to ordinary hydrogen is about the same as that found on Uranus and Neptune, as well as in Earth's water. This ratio is significantly higher than that which is likely to have occurred in the space between the stars, around the primordial Sun, or even near the giant planets Jupiter and Saturn. Not only does this fact support the idea that comets formed in the outer solar system beyond Saturn, but it also suggests that comets provided much of the Earth's water.

As Uranus and Neptune grew more massive, their gravitational pulls started to scatter the comets, much as they both hurled the Voyager 2 spacecraft out of the solar system after it passed Uranus in 1986 and Neptune 3 years later. Over hundreds of millions of years, this dispersal resulted in an expanding swarm as comets left the hive. Some four-fifths of these ancient comets were ejected from the solar system. Others were ejected directly by Uranus and Neptune into a spherical shell of comets located about a light year away from Earth; named after the Dutch astronomer Jan Oort, this "Oort cloud" remains as the main storehouse of comets in the solar system. Since their formation, gravitational perturbations, probably by nearby stars, have disturbed the paths of certain comets in the Oort cloud, sending them out into the space among the stars or inward toward the Sun. The Comet Levy that first appeared in the inner part of the solar system in 1990 is a likely example.

The remainder of the original comet swarm still within the solar system found their way towards Jupiter and Saturn, and many such comets went into Earth-crossing orbits and collided with Earth. Still other comets never got redistributed at all; these probably remain in the outer solar system as a group now called the Kuiper belt (after the late University of Arizona scientist Gerard Kuiper). The Kuiper Belt is believed to be the original storage ground for most "periodic comets," like Halley (see Figures 14.1 and 14.2), Encke, and Shoemaker–Levy 9, that orbit the Sun in under 200 years. Since 1992, astronomers have been discovering Kuiper Belt Objects, which are actual representatives of this ancient swarm. The present Oort cloud is more likely a sphere of comets that extends from perhaps 20 000 to beyond 150 000 astronomical units (an astronomical unit is the average distance between Earth and the Sun), or from 2 000 000 000 000 miles to 14 000 000 000 000 miles from the Sun.

How are comets made?

Most comets have three parts – a nucleus, coma, and tail. The nucleus is the one permanent feature of all comets. This tiny member of the solar system, almost never directly observed, generates some of the smallest and some of the largest objects in the solar system. The nucleus releases atoms, molecules, and particles of dust that can stretch out in a comet's tail for hundreds of millions of miles.

Figure 14.1. On December 18, 1985, using the 20-inch Tumamoc Hill reflector and the International Halley Watch CCD system, S. Larson, S. Tapia, and I took a series of 14 exposures of Halley's Comet (1P/Halley) totalling 960 seconds. The exposures were "co-added" to form this single image showing the jet structure to lower right. The star trail line to left is actually a recording of the changing position of the nearby star relative to the comet as it moved across the sky from picture to picture. Courtesy NASA/International Halley Watch.

The nucleus

Until the multiple spacecraft flybys of Halley's Comet in 1986 (see Plate II and Figure 2.1), a comet nucleus (visible as its central condensation, and also called its nuclear condensation) was believed to resemble what Harvard planetary scientist Fred Whipple called a "dirty snowball" – a ball of ices mixed with dusty particles. The spacecraft sent back data that indicate that Halley's nucleus has a much greater concentration of dust than expected, so

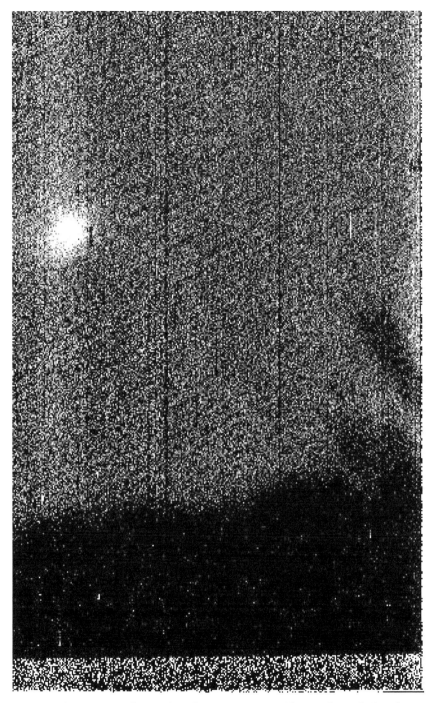

Figure 14.2. On the evening of January 27, 1986 (about 12 hours before the explosion of Shuttle Challenger) I was taking a series of CCD exposures of Halley's comet (1P/Halley) low in the western sky in twilight. This was a short test image. When it appeared on the screen, it looked so messy, with the comet off center and something at the bottom of the frame, that I reached for the delete key. In the nick

that it is really an icy ball of mud. The "snow" consists of a mixture of frozen water, and frozen gases including carbon dioxide (CO_2), hydrogen cyanide (HCN), and other gases containing carbon and sulfur. Some of these frozen gas molecules are believed to be mixed with or trapped within the water ice and dust.

As a comet moves toward the Sun, increasing temperature forces its nuclear ices to sublimate, or transform from solid directly to gas. In Halley's case, material equivalent to an average of about a meter's thickness, spread out around the comet, escapes each time it rounds the Sun. However, most of a comet's ejected material comes from several volcano-like vents in the nucleus, which are the primary source of the material in the coma, while the rest of the nucleus is quiescent. In addition to releasing material explosively from the nucleus, these vents can change the comet's orbit unpredictably.

The coma

Surrounding the nucleus is the **halo** of material called the coma. Ionized gas and dust is thrust out from the vents at thousands of feet per second carrying dust particles into the coma, but this activity does not usually become apparent until the comet is within three astronomical units, or about 300 million miles, of the Sun. An especially active comet, like Halley (see Figure 14.1, Plate II and Figure 2.1) or Hale–Bopp (see Plates VI and VII), has a detailed structure within the coma; **envelopes** or **hoods** are often seen concentrically placed around the central condensation.

The coma and nucleus together are referred to as the **head** of the comet. Surrounding the head is a huge cloud of atomic hydrogen gas emitting ultraviolet light. First observed in Comet Tago–Sato–Kosaka in 1969, this hydrogen envelope can stretch out over six million miles around the nucleus.

The tail

The Sun affects a comet's behavior by heating its ices, and also in other ways. Radiation from all parts of its electromagnetic spectrum, including infrared, visible, ultraviolet, and even X-ray radiation can affect the electric charge of the gases in the comet's coma and tail, ionizing them and forming (or increasing) a tail. A comet's tail has two major components: ionized gas and dust.

Caption for Figure 14.2. (cont.). of time I realized that the bottom was actually distant mountains and a foreground cactus plant. Thus, I was able to preserve the last known photograph of Halley's comet taken before it rounded the Sun 2 weeks later on February 9, 1986. Courtesy NASA/International Halley Watch.

The ion tail consists of molecules released by the nucleus that have been stripped of an electron, and thus become positively charged or ionized. This tail tends to be bluish. Once the tail begins, the direction and velocity of its particles are affected by the solar wind, which is a stream of electrically charged ions and electrons blowing out from the Sun. The solar wind carries magnetic fields which drag cometary ions along with them away from the Sun.

The dust tail is different. Particles of dust, made partly of silicate material and about one micrometer in size are carried into the coma by the ion tail. Once the particles leave the comet proper, they are pushed away from the Sun while at the same time moving with the comet's orbital motion. In this way, the comet dust tail becomes curved as dust spreads in the plane of the comet's orbit.

A comet's **anti-tail** is the part of the tail that, because of the observer's position on Earth, appears to be pointed in a direction opposite to that of the main tail; the main tail points away from the Sun, while the anti-tail appears to point toward it.

Families of comets

As a comet orbits the Sun, it could encounter the gravitational field of Jupiter or of one of the other major planets. When this happens, the planet has captured the comet and changes its orbit. Jupiter has captured many comets that have made at least one pass close enough so that the planet's gravity altered the comet orbit. These comets are said to belong to Jupiter's family. The periods of revolution about the Sun of Jupiter family comets average 6 or 7 years.

Sungrazing comets

As opposed to a family of comets, a group of comets includes multiple comets that share the same orbit. The Kreutz sungrazing comet group consists of hundreds of comets that pass very close to the Sun, often crashing into it. Included in this group are the Great Comet of 1882, Comet Pereyra in 1963, Comet Ikeya–Seki of 1965, and Comet White–Ortiz–Bolleli of 1970, not to mention hundreds of fainter comets found by the orbiting Solwind, Solar Maximum Mission, and SOHO observatories. Since these comets share a similar orbit, their brightness and grandeur depend on two factors: the size of the comet and the time of year that it reaches perihelion. Rounding the Sun in July, 1963, Comet Pereyra was bright but far from spectacular. There are two times during the year, in the periods from February 15 to April l, and from September 1 to October 15, that the sungrazing comet track is visible in the dark night sky. Both the 1882 and the 1965 comets appeared during these times and put on splendid displays. According to Brian Marsden, these two comets almost certainly split from a single body at their last perihelion, which probably was the comet of 1106.

NOTES

1. S. Drake and C. D. O'Malley, translation, *The Controversy on the Comets of 1618* (Philadelphia: University of Pennsylvania Press, 1960), 19.

15

Visual observing of comets

… He reads much,
He is a great observer, and he looks
Quite through the deeds of men.…

Caesar describing Cassius and other good comet observers in Shakespeare, *Julius Caesar*[1]

Have you ever wondered, when looking at a comet through a telescope, how big it really is? With some careful observations and simple calculations, you should be able to figure it out. Information like this is possible to obtain from your own backyard, and through your own telescope. Also, comets are exciting objects to observe because each is different. A new comet can appear anywhere, and it is interesting to follow as it moves across the sky. On a single night, a comet's physical appearance can be extraordinarily complex, but that appearance also changes over time. As the comet gets closer to the Sun, its coma becomes broader and more complicated. Its appearance can change also as the Earth–Sun–comet geometry changes. In 1986, as Comet Halley rounded the Sun, I watched its appearance change from week to week. Reacting more strongly to the Sun's heat, by the beginning of 1986 it was just starting to send out eruptions of dust called **jets**. These eruptions became more frequent until, on the evening of January 28, Halley was so close to the Sun that it was difficult to see it in the twilit western sky. A month later, the comet reappeared in the eastern dawn sky looking completely different: Its tail was much longer, its central part was far more condensed, and it was much brighter.

In the following weeks the comet continued to change its appearance as it edged farther from the Sun and drew closer to Earth. By early April it was swinging past Earth with such velocity that it looked different each night; the angle of the tail changed, and its coma seemed very active. Then in early May, the comet began to swing away from the Earth, and then its tail appeared to grow rapidly. On the night of May 2, 1986, its tail stretched more than 20 degrees across the sky, and its appearance was so striking that I telephoned my family, then vacationing in upstate New York, to go out to see it. I did not even have to provide directions.

Comets change in appearance for two reasons. First, the changing Earth–Sun–comet geometry means that the comet changes because the way we look at it alters. The second reason is that as the comet approaches and then leaves the Sun, it really does change its appearance as its activity level varies. When a comet is changing this dramatically, drawing its nightly changes in appearance is interesting and fun.

Drawing a comet

Good comet drawings have value even when compared with photographs or CCD images. The eye is very good at resolving fine detail, especially during moments of good seeing, and it can respond to an exceptionally wide range of intensity at one glance. A careful depiction of the relative intensity and position of the central condensation within the coma can yield valuable data on the state of activity of the nucleus or of "hot spots" – erupting jets or other anomalies – on its surface. Determination of the comet's rotation period and precession (wobble) rates are possible using drawings. Examples of drawings of different comet types and features are given in Figures 15.1 to 15.9. Such drawings can result in real contributions to understanding the behavior of comets.

A faint comet might not offer more than an undifferentiated, amorphous appearance. Examples of such comets are given in Figures 15.1 and 15.2.

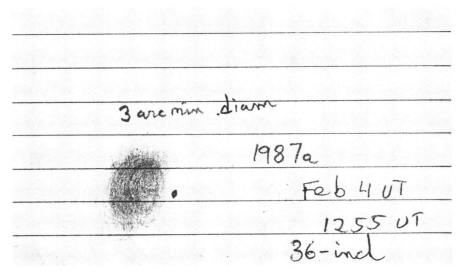

Figure 15.1. Comet Levy 1987a (C/1987 A1) drawn on February 4, 1987, through the 36-inch reflector atop Kitt Peak, in Arizona.

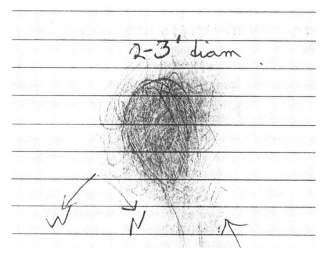

Figure 15.2. Comet Nishikawa–Takamizawa–Tago 1987c (C/1987 B2) drawn on January 25, 1987, through the 16-inch reflector at Jarnac Observatory, Arizona.

As a comet brightens, it is likely, although not always the case, that it will develop various features, which are shown in the drawings in Figures 15.3 to 15.9, and in the photograph in Figure 15.10.

- **Haloes** are round or oval enhancements surrounding all or part of the comet's central condensation (Figure 15.3).

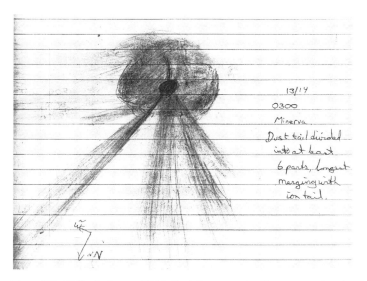

Figure 15.3. Halley's Comet (1P/Halley) drawn on April 13, 1986, showing halo and multiple dust tails, through the 6-inch reflector on board MS *Bucanero* in the Galapagos Islands.

- **Fans** are sectors of brighter material emanating from the central condensation (Figure 15.4).

Figure 15.4. Complex fan structure in the coma of Halley's Comet (1P/Halley) drawn on April 27, 1986, through the 16-inch reflector at Jarnac Observatory, Arizona.

- **Jets** are radial features emanating from the central condensation (Figure 15.5; see also Figure 14.1).

Figure 15.5. Halley's Comet (1P/Halley) drawn on April 12, 1986, through the 6-inch reflector on board MS *Bucanero* in the Galapagos Islands. The comet shows a large dust jet.

- **Envelopes** *or* **hoods** are features with more than one level of brightness surrounding the central condensation, exhibiting a discontinuous change in stepping to the next level (Figure 15.6).

Figure 15.6. Comet Hale–Bopp (C/1995 O1) drawn on the morning of March 11, 1997, through the 16-inch reflector at Jarnac Observatory, Arizona. The comet shows a very condensed nuclear condensation surrounded by a complex structure of envelopes.

- **Spines** are bright, sharp, narrow streaks leading from the central condensation into the tail (Figure 15.7).

Figure 15.7. Spine on Comet Takamizawa–Levy (C/1994 G1), drawn by the author the night after the comet's discovery (April 15, 1994) through the 16-inch reflector at Jarnac Observatory, Arizona.

- **Streamers** are soft-edged bright streaks seen in the coma and comet tail. The **"shadow" of the nucleus** is not really a shadow, but a dark streak leading from the central condensation into the tail. Both these features are shown in Figures 15.8 and 15.9.

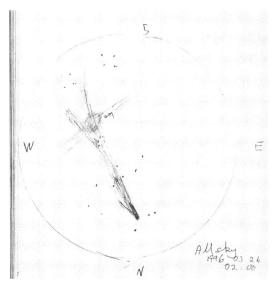

Figure 15.8. A once-in-a-lifetime memory: All-sky drawing showing Comet Hyakutake (C/1996 B2) on March 26, 1996; an unaided eye view from Jarnac Observatory, Arizona. The head of the comet is midway between Polaris and the two stars in the bowl of the Little Dipper (Ursa Minor, the Little Bear). The tail stretches through the Big Dipper (or Plough, in Ursa Major, the Great Bear), through the Gegenschein in Leo, and ends in the southern sky just east of Corvus.

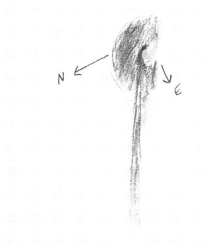

Figure 15.9. Comet Swift–Tuttle (109P/Swift–Tuttle) drawn on November 14, 1992, through the 16-inch reflector at Jarnac Observatory, Arizona.

- A comet's **anti-tail** is composed of sunlit particles around the comet; the anti-tail points toward the Sun instead of away from it and becomes visible when the Earth crosses the plane of the comet's orbit (Figure 15.10).

Figure 15.10. Comet Levy 1990c (C/1990 K1) with anti-tail, photographed on February 12, 1991, through the 18-inch Schmidt camera on Palomar Mountain, California, by Gene and Carolyn Shoemaker and the author.

The more care you take with your drawing, the more accurately your work will render the sizes, shapes, orientations, and positions with respect to the comet's nuclear condensation. As is obvious from my above sketches, the drawings do not need to be works of art. Do not rush. Soft lead pencils or charcoal drawing supplies and paper stumps and erasers for smudging and erasing, all available from stationery and art supply stores, should be used to make your drawings. I have also found that it is useful to do a series of drawings, made with various magnifications, or even through different types of telescopes, from wide-field reflectors to long focal length refractors.

For a drawing to be successful, it needs to be accompanied by three pieces of information:

(1) date and time;
(2) the scale, typically in minutes of arc per millimeter of sketch; and
(3) the orientation, showing where north and east are (indicating east is necessary because different designs of telescopes present the sky in different ways, some having east on the right, others on the left).

Two advanced measurements can also be made. These are:

(1) The vertex distance. This is the distance, in minutes of arc, from the comet's brightest part to the vertex of one of its envelopes.
(2) The semi-latus rectum. Two of these make the comet's pseudo-latus rectum, which is a measurement from the brightest point of the coma outward to the edge of the comet's envelope (this is perpendicular to the axis of the coma) on both sides. The measurement on each separate side is called a semi-latus rectum. These values can help with studies of how the comet's structure is changing with time. Fred Whipple, the twentieth century's leading expert on comets, considered the semi-latus recta important factors in understanding changing behavior of the comet nucleus. He pointed out that the two semi-latus recta may not always be equal, and recommended that they be identified and recorded separately.

Observing with filters

Sometimes a comet is more easily seen using a "Swan band" filter. These filters emphasize the cyanogen (CN) spectral line from a comet producing some gaseous emissions. They will cause such a comet to appear brighter, but will have little or no effect on a comet whose coma and tail is mostly dust. When making observations with such a filter, be sure to record the particulars.

How large is the coma?

By a simple measurement of the size of the coma in the sky, it is possible to use your own observations to determine the true size of the comet you are observing. This process begins with an estimate of how large its coma is. The star atlas method is a good one to use, but it will work only if the atlas you use is detailed enough to show the stars in the field of view of your telescope. Draw the comet among the stars in the field of view of your telescope, then compare your drawing with the chart and measure the size of the coma.

A potentially easier approach is the drift method. Estimate the size of the eyepiece field by shutting off the telescope drive and measuring the time, in seconds, that it takes for the comet to begin to leave the field of view to the time the comet's coma has completely left the field. Repeat this measurement a few times to improve accuracy, and take an average of the times. Then you can use the equation below

$$D_c = 0.25t \cos \delta, \text{ where}$$

D_c is the apparent coma diameter in arcminutes,
t is the time, in seconds, for the comet to leave the field of view (or cross the crosshair), and
δ is the declination of the comet in the sky. (The declination is important, since comets near the equator will appear to swing through the field more quickly than comets further north or south. In fact, this method loses much of its accuracy if the comet is north or south of 70 degrees declination.)

The drift method is more accurate if you use an eyepiece equipped with crosshairs. Make sure that the crosshairs are aligned east–west and north–south. To do that, turn the telescope drive off and watch as the comet, or a star, drifts across the field. Adjust the eyepiece so that one of the crosshairs is parallel to the star's motion. The other one will then be north and south. Next, measure the time, in seconds, that it takes for the comet to move across the crosshair.

Computing the actual size of the coma

The star atlas and drift methods will yield the apparent size of the comet's coma. To convert this to an actual size in miles or kilometers, you need to compare the apparent size of the coma with the comet's distance from Earth at the time of your observation.

Table 15.1. *DC numbers of comets*

DC number	Description
0	Totally diffuse coma with uniform brightness, no condensation toward the center.
3	Diffuse coma with brightness increasing gradually towards the center.
6	Coma shows definite intensity peak at center.
9	Coma appears stellar or sharp edged.

The equation for doing this is

$$D = 2\pi \, \Delta k D_c / 21600', \text{ where}$$

D is the true diameter of the comet's coma,
Δk is the comet's distance from Earth in astronomical units, and
D_c is the apparent coma diameter in arcminutes.[2]
The devisor, 21600 arcminutes, is 360 degrees around the sky.

The result will give the true size of the comet's coma in astronomical units (AU), where one AU is the mean distance between Earth and the Sun (93 000 000 miles), so you can easily convert the value to kilometers or miles. Say, for example, that D turns out to be a large 0.001 AU; multiply 93 000 000 by 0.001 to get a coma diameter of 93 000 miles.

Degree of condensation

This measurement, also called DC, provides a description of the coma's intensity profile; that is, the change in brightness with distance along a diameter through the coma. The concept of DC was apparently introduced by the British Astronomical Association in the 1950s.[3] DC numbers range from 0 (diffuse image, no condensation, flat, smooth profile) to 9 (star-like image with stellar intensity profile); see Table 15.1. Occasionally, comets develop a coma with a sharp edge like a planetary disk. A DC of 9 means that the comet is sharp-edged, or star-like, in appearance; experienced observers will generally rate a DC of 9 because the coma is not diffuse at all.

How you measure a comet's degree of condensation depends to some extent upon the telescope and eyepiece you are using. A comet might appear uncondensed (DC of 3) through an 8-inch Schmidt–Cassegrain telescope; but seen at the same time through a rich-field telescope with a wide field, it might appear more strongly condensed toward the center, perhaps with a DC of 6. It is a good idea always to use the same type of telescope when observing a comet's

DC, and it is certainly important to note the type of telescope used for each observation.

Tail measurements

The majesty of a comet is determined, more by the length of its tail than by any other factor. Halley's Comet was spectacular in 1910 because its tail was so long, and in 1996 Comet Hyakutake's tail stretched more than halfway across the night sky. You can estimate a comet's tail length with respect to different stars in the field of view of your telescope or binoculars, or with the naked eye.

In addition to the length of a tail, it is interesting to note the angle at which it leaves the comet's coma. This value, called "position angle" or PA, is measured with respect to due north: a tail pointing due north has a position angle of 0 degrees; a tail pointing due east is 90 degrees, one pointing due south is 180 degrees, and one pointing due west is 270 degrees. The "drift method" is a crude but effective way of measuring position angle. As with measuring the coma diameter, in this method, start by centering the comet (or a star) in the field. With the telescope drive off, watch the comet move westward in the eyepiece. Then estimate what time on an imaginary clock face the tail is pointing to; if west is at 12 o'clock, then north is at 3, east at 6, and south at 9. A more accurate method of determining position angle is to plot the position of the head and tail on a star atlas and then measure the PA with a protractor. This method can be accurate to about 5 degrees.

Bright comets often sport two tails, a gas tail and a dust tail. The position angles of gas tails, which are generally straight, are relatively easy to measure. The PA of a long, curved dust tail should be measured at its beginning, where it first leaves the coma, and again at various positions in the rest of the tail. A measurement of the distance from the nucleus should be included. All these notes can provide information about the curvature of the tail in space.

Notes on the detailed structure of a comet tail are always useful. If you notice any change in the tail, (such activity is far more likely in the gas tail, where ions shoot out at very high speeds), you should note this as well, along with the time of your observation.

Calculating the true minimum length of the tail

In the same way that you calculated the size of the comet's coma, you can try to determine the true length of the tail in space. Use the equation $D = 2\pi \Delta k D_c / 21600'$ to get a measure of the length of the tail. This is the same equation used to measure the coma: it works for measuring the length or diameter of any part of the comet. For comets with long tails, your answer could well turn out to be a strong fraction of an astronomical unit, or even more than an astronomical unit. However – and this is a real caveat – the real length of a comet tail is not that simple to determine, since the way the tail is projected against the sky must be taken into account. Also, if the tail is strongly curved it might appear shorter than it really is. The result you get is really the minimum length of the tail.

In his autobiography *Starlight Nights*, Leslie Peltier commented about using a set of figures on a printed page – "It gave me a most satisfying feeling," he wrote, "to reflect that from a maze of figures in a magazine I could point my telescope to a comet in the sky."[4] Using the information in this chapter, it seems possible to go further than that; from the printed page you can find a comet, and by careful drawing and measuring you can learn about the comet's size as it moves through space.

NOTES

1. *Julius Caesar*, 1.2.201–203.
2. Daniel Green, *Guide to Observing Comets*, a special publication of the *International Comet Quarterly* (January, 1997), 74.
3. Green, 75.
4. Leslie C. Peltier, *Starlight Nights: The Adventures of a Star-Gazer* (New York: Harper & Row, 1965), 227.

16

Estimating the magnitude of a comet

Brightness falls from the air …

Thomas Nashe, 1567–1601, In Time of Pestilence[1]

One of the most interesting aspects of observing a comet is estimating its brightness. In this chapter we explore the different methods by which an observer can compare the brightness (or magnitude) of a fuzzy, diffuse comet with the magnitudes of stars. In estimating the magnitude of a comet, you are really doing a form of visual photometry. By comparing the brightness of the comet concerned with the brightness of stars of known magnitudes, it is possible to obtain a record of the changing behavior of the comet over time.

Comets change in brightness for three reasons. The first is that a comet will brighten and fade as it approaches and recedes from the Sun. A second reason is that even as a comet is leaving the vicinity of the Sun, if it gets closer to Earth its brightness can increase. Finally, the comet's internal brightness might change as its surfaces are exposed to sunlight since this may cause eruptions of gas and dust. Several methods developed by different observers have been used over the years to measure cometary brightnesses. The magnitude of the entire coma (including the central condensation within it) of a comet is called M1; that of the nuclear condensation alone is M2.

Magnitude is represented by a logarithmic scale, on which the brightest star has a magnitude of −1.5, and the faintest visible star has a magnitude of 6; a decrease in magnitude one unit represents an increase in apparent brightness of about 2.5.

Comparing a comet with a galaxy or nebula

Since comets and galaxies and nebulae are all fuzzy, it is tempting to compare the brightness of one with that of either of the other two, using quoted magnitudes for the deep-sky objects. This is not a good method. The magnitudes for many non-stellar objects are themselves unreliable, and the nature of the light of a comet in our own solar system is so different from that of a distant galaxy, consisting of hundreds of billions of stars, that comparing their magnitudes does not usually give a reliable result. It is better to compare the brightness of a comet with that of a point source like an individual star.

Estimating the brightness of a variable star: The AAVSO method

Before beginning the process of comparing the brightness of a comet with that of a star, I recommend that you try a variable star or two. The field of variable stars is a fascinating one that I have enjoyed for decades. The easiest method to use to estimate the brightness of a variable star is that used by the American Association of Variable Star Observers (AAVSO).[2] Choose a star from their large series of star charts, and estimate its brightness in this manner:

(1) Choose two comparison stars on the chart, one a bit brighter, and the other a bit fainter than the variable star under investigation.

(2) Compare the brightness of the three stars, noting where the variable falls in between the two comparison stars. For example, if the magnitude of the brighter star is 7.0, that of the fainter one is 8.0, and that of the variable is 3/5th (or 6/10th) of the brightness of the brighter star, then its magnitude is 7.6. If the magnitude of the brighter star is 9.6, that of the fainter one is 10.3, and that of the star is 5/7th of the brightness of the brighter star, then its magnitude is 10.1.

(3) Beware of the Purkinje effect! Red stars have an effect on the eye in which they appear to get brighter as you watch them, even as seconds pass. To avoid this problem, look at a red star using quick glances.

Transferring this variable star approach to comets is not a straightforward process. Notwithstanding the problem of comparing the brightness of a fuzzy comet with that of a star, comets move across the sky, and the comparison stars you use might be far from the comet. The next four sections suggest different methods for estimating the magnitudes of comets.

The VSS or In–Out method

The most common approach is the VSS method. This is more popularly known as the Sidgwick method, but since others came up with it earlier than he did, it is now called the Vsekhsvyatskij–Steavenson–Sidgwick or more easily, the VSS method.[3] To estimate the brightness of the coma of a comet using the

VSS method, follow the steps given below. It helps if you have an atlas or program that lists star magnitudes; the AAVSO Star Atlas is an excellent source for star magnitudes all over the sky.[4]

(1) Study the coma until you are familiar with its "average" brightness. This is easy if the entire coma is uniform (has a low degree of condensation or DC) but not so easy if the comet has a highly developed central condensation.

(2) Using an atlas that incorporates star magnitudes, find a comparison star at about the same altitude and, if possible, in the same part of the sky as the comet.

(3) Throw the star out of focus so that it is the size of the in-focus coma.

(4) Compare the star's out-of-focus brightness with that of the in-focus coma.

(5) Repeat steps (2), (3), and (4) with more stars, until you have a star less than half a magnitude brighter than the coma, and a second one less than half a magnitude fainter than the coma.

(6) Interpolate between the two stars to assign the coma its magnitude.

The VBM or Out–Out method

This was formerly known as the Bobrovnikoff method, but since George Van Biesbroeck used it earlier, the *International Comet Quarterly* recommends that we now call it the Van Biesbroeck–Bobrovnikoff–Meisel or VBM method.[5] In this method, both the comet and the stars are thrown out of focus (defocused), as follows:

(1) Use a magnification of 1.5 to 2 power per centimeter of aperture (4 to 5 power per inch of aperture) to minimize the apparent size of the comet. Thus, if you are using an 8-inch telescope, use a magnification of between 32 and 40.

(2) Select several nearby comparison stars, some brighter and some fainter than the comet.

(3) Throw the eyepiece out of focus until the comet and stars have a similar apparent size. This is not as complicated as you might imagine, because the comet's apparent size will not change nearly as much as the sizes of the star images when defocused. This method, of course, does not work with a comet so faint that it disappears when defocused; in that case, use the VSS approach.

(4) Choose a brighter and a fainter star and interpolate the magnitude of the comet.

(5) Repeat step (3) with more star pairs.

(6) Average all the measurements you have taken in steps (3) and (4). Record the magnitude to the nearest tenth of a magnitude.

N.B.: This method does not work well if the comet is very diffuse or its coma is unusually large; the magnitude you get will likely be too faint.

The "Modified Out" method

Developed by comet observer Charles Morris and independently by Stephen J. O'Meara,[6] this method combines the "In–Out" and "Out–Out" methods into a single process. It matches the diameter of the moderately defocused comet with that of a defocused star. It is more complex than either of the methods that it is built on, but it is looked upon as more accurate by some seasoned comet observers. Follow these steps:

(1) Defocus the comet's coma until the surface brightness is approximately uniform.
(2) Remember the size of the image obtained in step (1).
(3) Match the comet image size with the image sizes of out-of-focus comparison stars. The stars will be more defocused than the comet.
(4) To estimate the comet's magnitude, compare the surface brightness of the defocused stars and the memorized comet image.
(5) Repeat the steps (1) to (4) a few times, averaging the result until an accurate estimate of magnitude (to the nearest tenth of a magnitude) is made.

The Beyer or Extrafocal Extinction method

The Beyer method is another approach to focusing comets that is useful only for observations in a dark sky. It works like this:

(1) Defocus the comet until it vanishes against the sky.
(2) Search for a star of known magnitude that disappears with the same amount of defocusing.
(3) Repeat steps (1) and (2), and take an average of the stars you have chosen to get the comet's magnitude.

Observing hints

If you are having trouble finding reliable comparison stars to estimate the magnitude of a comet, try the AAVSO (American Association of Variable Star Observers) charts for the stars around Polaris (see Figures 16.1 and 16.2).

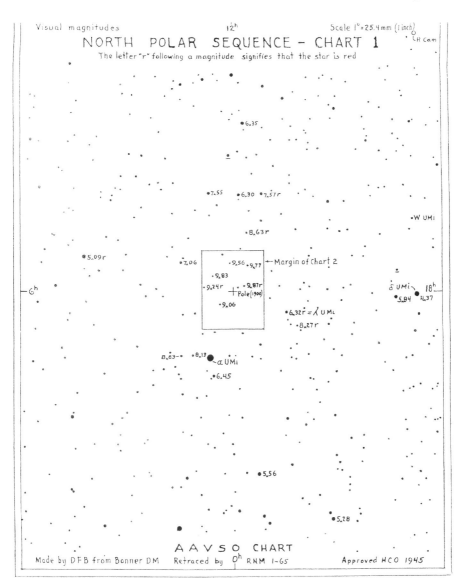

Figure 16.1. The north polar magnitude sequence for stars down to 10th magnitude. Courtesy AAVSO.

The advantage of this "north polar sequence" is that the stars are always visible, at least in the northern hemisphere, and offer a uniform sequence of stars that you can use with any faint comet.

It is always best to use the lowest magnification that allows you to see the comet easily. Also, do not use comet filters (such as the Swan band described in Chapter 15) when making estimates of magnitude.

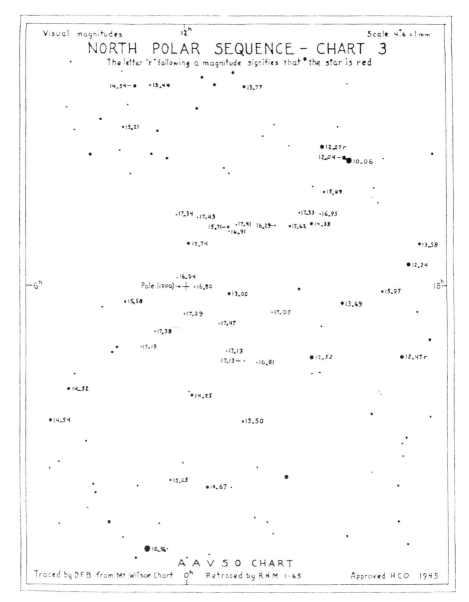

Figure 16.2. The north polar magnitude sequence for stars fainter than 10th magnitude. Courtesy AAVSO.

NOTES

1. *Oxford Dictionary of Quotations* (Oxford: Oxford University Press, 1955), 361.
2. A sampling of introductory star charts, with hints on estimating magnitudes, can be found in David H. Levy's *Observing Variable Stars: A Guide for the Beginner* (Cambridge: Cambridge University Press, 1989 (hardback) & 1998 (paperback)).

3. A discussion of methods of estimating magnitude can be found in: Daniel Green, *International Comet Quarterly*, 18 (October, 1996), 186; and in his *Guide to Observing Comets*, a special publication of the *International Comet Quarterly* (January, 1997), 63.

4. Charles E. Scovil, *The AAVSO Variable Star Atlas* (Cambridge, MA: American Association of Variable Star Observers, Second Edition, 1990).

5. Green (1996), 186.

6. Green (1996), 186, and (1997), 63.

17

Taking a picture of a comet

Comets then undoubtedly constitute a part of the great chain of nature;
but the singing of birds, the swarming of bees, the minutest atom that
floats in air, form likewise a part of this wonderful concatenation: and it
would be equally reasonable to consult them, as comets.

Peter L. M. de Maupertuis, March 26,1742[1]

Two hundred fifty years after the great European astronomer and
mathematician Maupertuis wrote these words, we do indeed consult comets,
not for their predictive power, of course, but to learn about them as fascinating
objects in space. Maupertuis would have been surprised and delighted to know
how that consultation takes place today – through the lens of a telescope, using
emulsioned film or electronic cameras.

We take pictures of comets to learn about how they behave, but also for
the sublime beauty of their artistry. The two most rewarding photographs I
have ever taken in my life had to do with comets. The first involved one of my
own comets, Levy 1990c (C/1990 K1); the second was my Comet Hyakutake
(C/1996 B2) cactus picture.

The Comet Levy photograph incident took place in early September 1990. It
was a clearish night in early September, 1990. I had set up my Schmidt camera
to take a picture of this bright comet that I had discovered a few months before.
Comet Levy was quite close to a bright star in Scorpius that night, and with the
waning Moon having just left the evening sky, I wanted to get a picture of it. The
sky was clear much of that afternoon, but as dusk settled and I loaded the film,
dense clouds formed and it actually started to rain. I leaned over the telescope to
cover it. Looking toward the southwest I could see the sky clearing near where
the comet was. The rain became a drizzle, then stopped. Meanwhile lightning
started to flash. I quickly removed the shutter, took a single 3-minute exposure, grabbed the film holder and put it in my pocket, closed the observatory,
and ran inside. The result is Figure 17.1.

Figure 17.1. Comet Levy 1990c (C/1990 K1) photographed with an 8-inch f/1.5 Schmidt camera during a thunderstorm, in early September, 1990, at Jarnac Observatory, Arizona.

Six years later, Yuji Hyakutake's discovery of a faint comet lurking in the southern sky caused a sensation for observers in the northern hemisphere. Comet Hyakutake was headed for a pass some 9000000 kilometers north of Earth in March. I expected a bright comet; what I did not expect was a bright comet with a tail that stretched across the entire sky. I also did not expect that it would be the cause of a very difficult photo session. On the night of Friday, March 15, Comet Hyakutake had brightened to the point that it was visible to the naked eye. On that evening I was visiting the home of my friend Tim Hunter. As I looked toward the south I could see the fuzzy head and tail of Comet Hyakutake. I took one step to get a better view, and then another. It was one step too many. Toppling off the front step, I fell to the ground and sprained my ankle. The comet was brightening fast, but I had to observe the entire apparition in a cast and on crutches. In the nights after my fall, I watched Comet Hyakutake brighten as its pushed its northward course, and its tail grew longer. By the time it reached Ursa Major it was the Great Comet of 1996, with a tail that stretched most of the way across the sky. For the three nights starting on Sunday, March 24, Hyakutake was the most dramatic comet I had ever seen.

On the morning of March 26 (Session **9663 AN), Jim Scotti drove me a few miles west of my home to a strand of beautiful saguaro cactus plants. We found a couple of particularly photogenic ones, and we set our Yashica twin-lens cameras up. Leaning on the crutches, I centered my camera on the comet and took the photo that is Figure 17.2.

Considering the hundreds of films of known comets and search films covering much of the sky that I have taken over the years, when I sat down to write something on astrophotography, the memories came flooding back. The moral: Photography is a hard teacher. There is a lot to remember, both in setting up exposures, guiding them accurately, and then developing them carefully, but when the picture is done, it offers a reminiscence that cannot be surpassed.

Which films, cameras, and exposure times are best for comet photography? In the rest of this chapter I introduce the exciting and challenging world of comet photography.

Three types of comet photography

Comets offer at least three types of goals for photographers. You can take a wide-angle picture that includes the comet's head and tail, or photograph just the coma, or the nuclear region.

Wide-angle photography

This is the easiest form of comet photography and begins with setting up any camera fitted with a time exposure setting onto a tripod (or leaning it against a wall, as I have done), pointing it at the comet, and exposing for 1–5 minutes. The comet and the star images will be trailed, but the image that results can still be a fine one.

Using color filters can enhance these photographs. For example taking successive images, with no filter, a blue filter, then an orange filter, will bring out the different dust and ion tails of the comet. These filters, made of glass or gelatin, can be simply placed in front of the lens. Kodak gelatin filter No. 47A is a recommended blue filter that should bring out the ion tail, while filter No. 21 works as an orange filter that will emphasize the dust tail. The pictures

Figure 17.2. Comet Hyakutake (C/1996 B2) as seen behind a saguaro cactus plant, March, 1996. Photograph by the author, using a Yashika twin-lens camera, in Arizona.

can then be combined either in a darkroom or, if they are scanned into a computer, using a photographic computer program.

Moderate-scale photography

Most comets – the vast majority in fact – present a small, somewhat elongated coma. Through a telescope, it is possible to take good pictures of this coma, which might show features that are undetectable visually. If the comet is a major one with a complex structure, photographing the coma region will reveal the wonderful features that often mark the beginning of the tail as it leaves the coma. The filters recommended in the previous section will also be good for moderate-scale photographic efforts.

Near-nucleus photography

When a comet is particularly active, it is useful to try some high-resolution imaging of the area around its nucleus. I suggest that you use a telescope of focal length 1500 mm or more. Because no two comets are identical either in brightness or in complexity of the region around the nucleus, I recommend that you take a series of bracketed exposures: start with a short exposure, double it, and then keep on doubling it. The chances are that one or two of your exposures will be ideal, but even the ones that are not might reveal other types of detail, including jets and additional features, or even the possibility of fragments splitting off the nucleus.

Films

The choice of film for comet photography really applies to any kind of astrophotography. There are almost as many choices of favorite films as there are photographers. Use of the finest-grain film available, like Kodak's Technical Pan 2415 or 4415, is important if you want to capture the most detail. However, this film is very slow unless it is hypersensitized or "hypered" (see Chapter 7 for a discussion on this process). If you do not have the equipment to hyper, you can try Kodak's T-Max film, a moderately fast film (two speeds are offered, 100 and 400) that is also quite fine-grained; I have found that this film works quite well for comets. Also, Kodak's Plus X pan or Ilford's FP4 are good films of moderate speed. Kodak's Tri-X, is a fast film, but is so coarse-grained that the image quality suffers. For photofinishing, I recommend a high-contrast developer like Kodak's D-19, although I have found that Kodak's HC-110 also works well, especially if you are not developing often, and the concentrate lasts almost indefinitely. I have also had good results with T-Max developer. I use Ilford and Kodak rapid fixers.

In color photography, two films seem to stand out: The new Kodak Elite Chrome 100 Extra Color (E100VS), and Kodak Pro-400 color negative film. Both have rewarded their users with good results. If you prefer slides, Kodak's P1600 Ektachrome is highly recommended, although I have also had good results with the full line of Ektachrome films.

If you use a CCD camera, of course, you need not worry about film or chemicals at all. Chapter 8 contains a discussion of photographing using CCDs.

How to guide an exposure

Most comets move across the sky at some large fraction of a degree per day. To capture the details of this moving object, it is important to guide on the center of the comet, rather than on a nearby star. The easiest method to accomplish this is to guide directly on the central condensation within the coma. An illuminated crosshair eyepiece will help in keeping the condensation centered. If the coma has no discrete condensation, you can try to keep the same part of the coma centered in the eyepiece.

The best general advice that a photographer can give is to keep taking pictures. The more you take, the larger your collection of satisfying images will become. The pictures you take will become cherished memories. Plate VIII is one of these. I set my Yashica twin-lens on a tripod pointing towards Comet Hyakutake, opened the shutter, and went out visual observing with Jim Scotti. As dawn approached I drove him home, but as I started the return trip I realized that the Yashica was still gathering photos of the rapidly moving comet. I hurried home just in time to turn off this completely unguided exposure before dawn began. The result is a very long series of star trails, with the comet's own motion across the sky – different from the diurnal motion of the stars – recorded over 90 minutes. This "mistake" turned out to be a photograph I will always cherish.

NOTES

1. Peter L. M. de Maupertuis, *An Essay Towards a History of the Principal Comets That Have Appeared Since the Year 1742*, translation (Glasgow: Robert Urie, 1770), 11.

18

Measuring where a comet is in the sky

The elements of a comet are the five articles which determine the position and magnitude of the parabola it describes, and which constitute its theory; namely, its node, inclination, place of its perihelion, perihelion distance, which is the square of the parameter, and the time when the comet arrives at its perihelion....

In the preceding letter [from Dr. Maupertuis] it has been remarked that Dr. Halley upon the Newtonian theory, had determined the elements of twenty-four comets. At present the number of those that have been accurately observed, and whose orbits are calculated, has more than doubled.

Peter L. M. de Maupertuis[1]

When anything new appears in the sky – a comet, an asteroid, a nova, or a supernova – the first thing the astronomical community needs to know, as accurately as possible, is its "precise position". This astrometric position, a measure of the comet's place in relation to the stars that apparently surround it in the sky, is a crucial first step to any further learning that we can accomplish from this newly found world in space. A series of astrometric positions will allow us first to learn where the comet is headed, so that we can continue to follow it. The positions allow us to compute an orbit that predicts its future behavior and lets us understand its past behavior, quite often, once the orbit is calculated from the positions, we can locate prediscovery positions of the comet a few months – or sometimes years – back in time. In 1770, orbits for some 50 comets were known; now in far greater detail, we know the orbits for some 1500 comets and tens of thousands of asteroids.

I have already told (see Chapter 6) of the astrometric story of Comet Levy 1988e (C/1988 F1) – of how Gene and Carolyn Shoemaker measured the comet's precise position from their photographs, and supplied accurate positions of the comet. Two months later, while trying to photograph the comet

again, the Shoemaker team discovered Comet Shoemaker–Holt. It was the astrometry, taken by observers all over the world, of both comets that allowed the
unraveling of the tale of a single comet that split apart some twelve thousand
years ago.

Basic astrometry

A precise position, accurate to a tenth of a second of arc in right ascension, and a second of arc in declination, cannot be obtained from visual observations. This accuracy can, however, be derived from photographs or CCD
images. The procedure of measuring these positions is called astrometry. With
the dizzying array of electronic telescopes and CCD systems now being made
available to the amateur community, astrometry is becoming a vital activity for
amateur astronomers.

In the old days of photography, measurements were made with a "measuring engine" – not a gasoline-spewing, piston-clunking monster but a large machine with wheels and screws, dials, gears, and scales. The machine had to be
free of "periodic errors" in its screws, and also free of play; measurers were advised always to measure the object and the surrounding stars from the same
direction. Experienced users of measuring engines were good at finding tricks
to improve the accuracy of their measurements. One advantage of the CCD
revolution is that measuring is done without any hardware like gears and
screws, and so is far less subject to personal errors. When the Shoemakers and I
were observing at Palomar, we used an apparatus I called Madame Guillotine;
this would produce a single measurement after about an hour of work involv-
ing the careful selection stars and the recording of their x and y coordinates
by hand. We would then use an astrometry program to analyze the results and
produce a precise position. Compared with that, twenty-first century astrometry programs come up with star selection, measurement, and a precise position every second. After spending huge amounts of time on the guillotine, you
can imagine what a pleasure I find using programs like PinPoint that measure
thousands of times more quickly and much more accurately![2]

Observing the object

Without a good observation, of course, there can be no astrometry, so
the first step in the process is to photograph the object using a CCD system. The
power of electronic chips is so great these days that an exposure of a minute
or two is usually sufficient (at Jarnac we use 90 seconds). The exposure needs
to be long enough so that the object will reveal itself on your CCD image. I

recommend that you try several exposures at differing lengths until you have one that clearly records the comet's center but ignores much of the surrounding coma and tail. This way, you have a clear center of the comet to plot, which should allow you to get a good position.

Whether doing it the old way or with CCDs, at least three stars, and ideally, eight or more stars, need to be measured in addition to the object under study. As Werner Heisenberg pointed out in his uncertainty principle, we cannot know with certainty something like the position of an object in the sky, but the more star positions we relate it to, the better our understanding of its position should be. If you are using photographs and a classical measuring engine, try these suggestions:

(1) Choose stars that are evenly distributed around the object.
(2) Measure the object during two stages of your process, once near the beginning, and again near the end.
(3) Measure the series of stars, and the object, first in one orientation, then rotate 180 degrees and measure everything again. Then take the average of the two sets to use in your reduction.
(4) Always have the measuring engine screw traveling in the same direction.
(5) After you have recorded the x and y coordinates of a star, go back and do the measurement a second and then a third time if the numbers in either axis are discordant. The number you record should be the average.
(6) After reducing your data, either you or your program can eliminate the data from some stars whose catalogue positions are poor (the program should recognize that the star's measured positions do not "fit"), but your final measurement should be accompanied by at least five stars.

Using CCDs

The first observing group consistently to use CCD imagery to do astrometry was the Spacewatch project at the University of Arizona. In the 1980s, Tom Gehrels, Jim Scotti, Robert Jedicke, David Rabinowitz, and other observers, developed the techniques of CCD astrometry. A decade later, professional and amateur astronomers throughout the world were teaming up to observe newly discovered comets and asteroids, and submit precise positions. The key to consistently good measurements is consistently good images; they need to be well guided and well focused. Then, astrometric engines like Pin-Point use software to do what the old measuring engines did. This program maps and provides astrometric solutions for entire images. To set it up, you need to provide information about the scale of your image, and the faintest star your telescope can detect. Once you have accomplished that, and the program understands the scale and magnitude limit of your image, it does the rest. Use the mouse to locate and click on the brightest pixel of the object you are measuring. This pixel normally, but not always, marks the place that will give the most accurate measurement. A dust jet erupting from the nucleus of an active comet could actually appear brighter than the nucleus; in that case you would be mistakenly measuring the brightest point of the jet and not the comet's center.

The elements of a comet orbit

When at least three precise positions have been obtained, it is possible to calculate the comet's orbit. The more positions and the longer the time interval among these positions, the better the estimate of the six orbital elements needed to describe a comet's journey through the solar system. The following are the six basic orbital elements from which it is possible to determine where a comet will be at a particular time.

(1) The comet's perihelion date, T: The date and time that it is closest to the Sun.
(2) The distance to the Sun at perihelion: This factor, q, is expressed in terms of an astronomical unit – approximately the average distance between Earth and the Sun.
(3) The eccentricity, e: If the orbit were a perfect circle, e is 0; if it were a parabola, e would be 1. The earliest orbits for most comets are figured as if e were equal to 1. If the comet turns out to be periodic, traveling in an ellipse, it will quickly deviate from the predicted path in the days and weeks after discovery.
(4) The inclination of the comet to the plane of the ecliptic: This angle of inclination is called i. Most periodic comets have inclinations less than 20 degrees. Long-period comets can have any inclination. Inclinations of 90 to 180 degrees refer to comets moving "retrograde" with respect to those with inclinations from 90 to 0 degrees.
(5) The longitude of the ascending node, Ω: This is the angular distance between the vernal equinox and the point, which we call the ascending node, where the comet crosses the plane of the Earth's orbit as the comet moves northwards.
(6) The argument of perihelion ω: This is the separation, measured again as an angle, between the ascending node and the point where the comet is closest to the Sun.

Finally, all these elements are accurate only at one time, which we call the epoch. Orbits can be computed for other epochs of interest to observers.

By doing astrometry on a comet, you are truly a part of the advance of knowledge. The observation you get, and the position you provide, will within a few days or hours result in an improved understanding of the orbit. What a way to contribute to science!

NOTES

1. Peter L. M. de Maupertuis, *An Essay Towards a History of the Principal Comets That Have Appeared Since the Year 1742*, translation (Glasgow: Robert Urie, 1770).
2. PinPoint's web site is http://dc3.com/.

PART V

Closing notes

19

My passion for comets

In the mean time, you will, I hope, be satisfied with knowing that this
comet has passed from Antinous into the Swan, and from the Swan into
Cepheus, with such rapidity, that it sometimes ran six degrees in
twenty-four hours. It proceeds towards the pole, and is not above ten
degrees from it. But it abates of its speed; and its light and tail are so
diminished, that we perceive it moves from the earth; and that for this
time, we have nothing either to hope, or to fear from it.

Peter L. M. de Maupertuis, regarding the comet of 1742, from Paris, March 26, 1742[1]

Only 90 minutes of darkness remain as I crawl out of bed, walk the
hundred or so feet to my observatory, roll open its roof, and turn Miranda, my
16-inch reflector, to the east. The sky has darkened a lot since I left it a few hours
ago. Save for distant coyote calls and an occasional chirping of a cactus wren,
the world around me sleeps. The night is still, poetic, and gorgeous. Setting
up the telescope's encoder system takes a few minutes, but does nothing to
spoil the artistry of this moment. Soon I am ready to go, and I begin searching
in a pattern of sky, at the rate of about a field per second, from north to south
near the eastern horizon.

First pass of the night

As I search, my mind goes back to earlier times, particularly to a period of time rich in comet sightings and lore. The period in question, the
1740s, comes to life this morning from an old book, *An Essay towards a History
of Comets*, that I had been reading.[2] Those were the times of discoverers like
Dirk Klinkenberg from The Hague and Philippe de Cheseaux of Lausanne, and
their remarkable comet of 1744. Between March 6 and 9 of that year, the comet
sported an overwhelming fanned tail with as many as 11 rays.[3]

As I continue to search, I imagine how much in the news comets must have been in the early 1740s. At its first appearance, Philippe de Cheseaux wrote of the Comet of 1744:

> it had no tail, at least perceptible to the naked eye; but in approaching the Sun, it acquired one which increased every day till it arrived at its perihelion; so that February the 17th, it was forty degrees long, and it still augmented considerably after the perihelion; for although the body of the comet could no longer be seen, the tail was visible two hours before sun rise … divided into five large streams, or bands…[4]

I look away from my telescope for a moment, staring at the eastern horizon and wondering what a sight that comet must have been two and a half centuries ago; and it followed another great comet by only 2 years! The 1742 comet excited the world of its time much as Comet Hyakutake excited us in our time. Discovered from the Cape of Good Hope, this 1742 comet rounded the Sun on February 8, then made a pass, much like Hyakutake, close to the pole in the northern sky.[5] I love eavesdropping on the letter that opens my old book, quoted at the beginning of this chapter and written by the celebrated mathematician and astronomer Peter Louis Morceau de Maupertuis to the Marchioness du Chatelet, as he tried to put the appearance of this comet into the perspectives of his time:

> You desire my opinion, madam, concerning the comet which is at present the general topic of conversation throughout Paris; and your wishes are to me commands. But what can I say to you of this star?… comets employ a much longer time than planets, in finishing their revolutions round the Sun. The slowest planet, Saturn, compleats his course in 30 years; while the swiftest comet employs 75 years for his, and it is highly probable that the greatest number are many ages performing their courses.[6]

Maupertuis's letter was a fascinating tour of ideas about comets, and their role in human affairs. He tried to assuage the good marchioness's fears of a comet being a disaster, or a bad star, but he correctly noted that comets could provide some real problems. He wrote:

> If a great comet should advance too near the earth, it might force it from its orbit, and oblige it to perform its future revolutions round the comet; wholly subjecting it either by its attraction, or, if I dare use the word, by involving it in its vortex: the earth, thus become a satellite to the comet, would be carried by it into those distant regions which it visits. Wretched condition for a free born planet, which has so long enjoyed a temperate sky! In short, the comet might in like manner rob us of our moon; and

were we to escape upon such easy terms, we might think ourselves very well off. But the most violent accident that could befal us, would be the percussion, or shock of a comet against our globe, which might break forth itself and the earth into a thousand pieces. In such case, doubtless, both these bodies would be destroyed; and gravity would immediately form one or many planets out of them. If the earth has never yet undergone these catastrophes, it has doubtless experienced many great changes. The prints of fishes, and even petrified fishes, which we find on places very different from the sea on the very summit of the highest mountains, are incontestable medals of some one of these events.

A less violent shock, such as would not entirely break our planet, would certainly cause great changes in the situation of lands and seas; the waters during such an accident would be greatly raised in some parts, and would drown vast regions of the surface of the earth, from which they would afterwards subside. To such a shock as this Dr. Halley attributes the deluge....

It is evident that, whatever may befal the earth, the other planets are equally liable to; unless the enormous size of Jupiter and Saturn should be their protection from the insults of comets. It would be a very curious sight for us to behold a comet approach, and fall upon Mars, Venus, or Mercury, and either break it into pieces in our view, or violently carry it away, in order to make it a satellite.

Comets may even attempt the Sun himself, and though they have not sufficient power to draw him after them, yet their magnitude and near approach might enable them to remove him from his present place. But Newton secures us from such a removal, by a conjecture drawn from the known analogy between comets and planets. Amongst these last, the least are situate nearest the Sun, and the greatest are most distant.[7]

What a prescient mind Maupertuis had! He correctly forecast not the details of cometary collisions but the fact they could happen; in so doing he admitted the possibility of a comet impact on Jupiter, such as we have recently witnessed, and the onslaught of comets into the Sun, as SOHO sees every week. Had he known that a comet nucleus is the size of a village, not a planet, he might have toned back his voice of doomsday.

Second pass of the night

Still at my telescope, I have completed my first pass in a small up-and-down sweep that gradually moves from north to south. I return north, begin searching an area closer to the Sun, and return to my reverie. Who else, over time, shared this passion for the cometary phantoms of the celestial opera? A

100 centuries ago, an early American native might have gazed at some long-departed comet from the very spot on which I now sit, and more than 100 years ago, the passion for comets was shared by people like Jean-Louis Pons, whose record has at least 27 discoveries, and George P. Bond, who observed from Harvard College Observatory as Donati's Comet completed its magnificent visit in 1857.

The passion continues through the late nineteenth century, when Amedée Guillemin published *The World of Comets*. This book contains a delightful idea by Andrew Oliver who, in 1772, two years after the Maupertuis book I have been reading was published "suggested that the purpose of a comet's tail, generated as the comet approaches the heat of the Sun, is to keep the temperature of the comet itself cooler for its inhabitants, and that as the comet recedes from the Sun, the tail and coma enwrap it, thus keeping the inhabitants warmer."[8]

In each age, comet enthusiasts were amazed at how much their culture had learned about comets from the previous one; so it is with our special time. Today we have the benefit of multiple spacecraft encounters with comets, a comet collision with Jupiter, and comets being found in record numbers. In a sense, we are so far ahead of those earlier people, but in a greater sense we are not ahead of them at all. Maupertuis died in July, 1759, just a few months before Halley's Comet announced its presence to confirm the great astronomer's prediction. I saw Halley in 1986, as part of the most ambitious observing program ever mounted for a single comet, the International Halley Watch.[9] Included in its massive, 26-CD archive are the data from ground-based observatories and amateur astronomers, as well as from the five spacecraft that studied the comet. Three of these craft, the Soviet Vega missions and the European Space Agency's Giotto, made very close passes by the comet and took *in situ* measurements and photographs. Halley, Maupertuis, and all the other people who have felt the passion for comets, would have been astounded.

Dawn's early light: Third pass of the night

With the first tinge of dawn appearing in the northeast, I swing Miranda back north again, lower the observing chair, and begin a final pass right at the horizon. It was during such a pass that one night during the summer of 2000 I independently discovered Comet 2P/Encke as it made one of its many returns about the Sun since observers first picked it up at the close of the 18th century. When I spotted that little comet, I felt that I was part of a chain of people who, throughout history, has identified with this visitor from space that comes by every 3.3 years. It is quite a chain: On January 17, 1786, the French comet hunter Pierre Méchain found the comet at fifth magnitude.

Figure 19.1. Damien Lemay took this picture – or rather these pictures of Comet Hale–Bopp (C/1995 O1), as a result of fatigue and excitement. In a letter to the author in July, 1997, he said:

> I have tracked Hale–Bopp more than any other comet. In April and May 1997, after a long stretch of clear weather, the call of the Comet had used much of my capability with the result that my body and soul became less reliable. On one particular night, after observing I accidentally processed a color slide film with chemicals for black and white negatives. The black and white 2415 hypered film which I was supposed to process, but didn't, returned to the freezer and eventually to my Schmidt camera, with the result of a double print.
>
> The exposures were taken respectively with a 55-mm lens and my 5.5-inch Schmidt camera, on the nights of April 27 and May 2. The first one was unguided (produced star trails) while the second was clock-driven in my observatory. While the star trails are obvious, a close look reveals the point stars of the guided Schmidt exposure.

Photograph courtesy Damien Lemay.

Nine years later, on October 20, 1805, Caroline Herschel discovered the comet as it returned. On October 20, 1805, Pons discovered it. Meanwhile, when a German mathematician named Johann Franz Encke calculated an orbit for this comet, he was surprised to suggest that it might be a comet that returns every 12 years. On November 26, 1818, Pons once again discovered this comet. Encke then connected Pons's new comet with the 1805 comet, but in so doing he shortened its period from 12 years to 3.3 and connected it with the discoveries of Méchain and Herschel.

Since then many people have recovered or observed Encke, including Horace Tuttle in 1875. This famous comet discoverer had codiscovered Comets Swift–Tuttle and Tempel–Tuttle, the parent comets of both the Perseid and Leonid meteor showers; but at the time of his recovery of Encke on January 25 that year, Tuttle was 3 days into a U.S. Navy court martial on charges of stealing more than $8000 from the U.S. Navy. As already related in Chapter 4, 3 weeks after Tuttle recovered Encke's Comet, the court found him guilty, but he received a light sentence, signed off by President Grant, of a dishonorable discharge from the Navy but no prison time.[10] One speculates that Tuttle's fame with comets helped get him off.

Stories like this one show that the saga of comets is enriched by its human side. As I made my own "recovery" of a faint Encke's comet that morning, I was thankful that I did not have $8000 to account for; but in a sense I was right there with Tuttle as he made his find during a terribly difficult period of his life. That human side of comets is something I think of now as I struggle to see the stars as they disappear into a brightening sky.

The hour is up; the search is ended for the night. I return to bed as Wendee lazily turns over, opens an eye, and mumbles with a smile, "Well?"

"No new comets tonight," I whisper with resignation. "But Wendee, what a sky! So clear, so sparkling – spectacular beyond belief." As we go back to sleep, I close my eyes with two images that I will cherish always – Wendee's sleepy smile, and the sight of the Milky Way straddling overhead in a quiet, ethereal star-filled sky, in that last quiet hour before the end of night.

NOTES

1. Peter L. M. de Maupertuis, *An Essay Towards a History of the Principal Comets That Have Appeared Since the Year 1742*, translation (Glasgow: Robert Urie, 1770), 43.
2. Maupertuis.
3. Gary W. Kronk, *Comets: A Descriptive Catalog* (Hillside, NJ: Enslow Publishers, 1984), 16–17.
4. Maupertuis, 52.
5. Kronk, 16.
6. Maupertuis, 21.
7. Maupertuis, 33–36.
8. Amedée Guillemin, *The World of Comets*, translated and edited by James Glaisher (London: Sampson, Low, 1877.)
9. For archive information, see http://pdssbn.astro.umd.edu/sbnhtml/comets/IHW/
10. The basis for this story comes from an unpublished memoir from Richard E. Schmidt, U.S. Naval Observatory.

Appendix
Levy's catalog of comet masqueraders and other interesting objects

This is a catalog of objects that have caused me to stop, look, and wonder during my 37 years of comet hunting. The asterisked objects, like 2, 14, 28, 36, 43, 45, and 58, and especially 37, 43, and 78, clearly "masquerade" as comets, especially when they are low in the sky where I usually spot them. The rest of the list contains different types of interesting objects – like no. L61, the Cetus Ring – that have stopped my search over some 2758 hours with my eye at the eyepiece (as of September, 2002).

Although the list is generally in chronological order of when I first spotted an object, several objects near the end were added later. Names in quotations are names that Wendee or I applied to specific objects.

| | | | (2000.0) Coordinates | | |
| | | | α (Right Ascension) | δ (Declination) | |
No.	Other designation	Spotted on			Comments[a]
L1	NGC1931	1/1/1966	05 31.4	+34 15	very condensed
*L2	NGC5457 M101	7/4/1966	14 03.2	+54 21	only visible under excellent conditions
L3	NGC6341 M92	7/5/1966	17 17.1	+43 08	thick well-defined nucleus
L4	NGC6254 M10	7/13/1966	16 57.1	−04 06	
L5	NGC5676	7/13/1966	14 32.8	+49 28	small ellipse, insignificant
L6	NGC6229	7/14/1966	16 47.0	+47 32	fairly bright, compact globular cluster
L7	NGC5055 M63	7/14/1966	13 15.8	+42 02	sunflower galaxy, fairly distinct but faint
L8	NGC1624	7/16/1966	04 40.4	+50 27	cluster with nebulosity; rich background
L9	NGC2403	7/16/1966	07 36.9	+65 36	galaxy, bright and striking

No.	Other designation	Spotted on	(2000.0) Coordinates		Comments[a]
			α (Right Ascension)	δ (Declination)	
L10	NGC2655	7/20/1966	08 55.6	+78 13	galaxy, fairly distinct
L11	NGC4605	8/5/1966	12 40.0	+61 37	"long"
L12	NGC7078 M15	8/23/1966	21 30.0	+12 10	a favorite globular cluster
L13	NGC6720 M57	9/7/1966	18 53.6	+33 02	ring nebula
*L14	NGC2068 M78	9/10/1966 8/16/2002	05 46.7	+00 03	comet-like nebula complex includes NGC2071 around star to north
L15	NGC4826 M64	9/19/1966	12 56.7	+21 41	black eye galaxy; elongated and easy to see; dust lane
L16	NGC3627 M66	3/3/1967	11 20.2	+12 59	easily visible galaxy
L17	NGC6853 M27	3/3/1967	19 59.6	+22 43	dumbell planetary nebula
L18	NGC2392	5/5/1967	07 29.2	+20 55	clown face planetary nebula in Gemini
L19	NGC6207	5/6/1967	16 43.1	+36 50	galaxy close to M13
L20	NGC6402 M14	5/6/1967	17 37.6	−03 15	distinct globular cluster
L21	NGC6838 M71	1967	19 53.8	+18 47	distinct cluster
L22	NGC3034 M82	12/24/1967	09 55.8	+69 41	rare, long irregular galaxy
L23	NGC5024 M53	5/5/1968	13 12.9	+18 10	globular cluster
L24	NGC3031 M81	11/27/1968	09 55.6	+69 04	galaxy in same field as 82 but brighter and larger
L25	NGC1904 M79	1/22/1969	05 24.5	−24 33	rare winter globular, small and distinct
L26	NGC4258 M106	6/25/1970	12 19.0	+47 18	pear-shaped galaxy
L27	NGC5377	8/6/1970	13 56.3	+47 14	galaxy just off Dipper handle
*L28	NGC5473	8/6/1970	14 04.7	+54 54	galaxy, comet-like, near M101
L29	NGC5474	8/6/1970	14 05.0	+53 40	galaxy, near M101
L30	NGC5866 M102?	8/8/1970	15 06.5	+55 46	galaxy
L31	NGC514	8/29/1970	01 24.1	+12 55	fairly easy galaxy under good sky
L32	NGC7331	9/3/1970	22 37.1	+34 25	indistinct but definite from city sky. Stefan's quintet nearby
L33	NGC1952 M1	8/29/1970	05 34.5	+22 01	bright super nova remnant
L34	NGC488	9/12/1970	01 21.8	+05 15	elongated galaxy
L35	NGC2420	9/12/1970	07 38.5	+21 34	cluster; fuzzy at first, distinct cluster on later examination

No.	Other designation	Spotted on	(2000.0) Coordinates		Comments[a]
			α (Right Ascension)	δ (Declination)	
*L36	NGC6364	1/11/1983	17 24.5	+29 24	comet-like galaxy in Hercules
*L37	NGC3055	1981	09 55.3	+04 16	"Wendee's galaxy." Looks like a diffuse comet
L38	NGC7753	4/28/1985	23 47.1	+29 20	slightly elongated galaxy, broad toward center
L39	NGC6791	date unknown	19 20.7	+37 51	faint open cluster
L40	NGC5466	date unknown	14 05.5	+28 32	loose globular cluster, diam. 10′
L41	NGC3810	date unknown	11 41.0	+11 28	round galaxy, bright core, diffuse
L42	NGC4340	date unknown	12 23.6	+16 43	galaxy in Coma Berenices
*L43	NGC7793	1/23/1989 12/24/1999	23 57.8	−32 35	like a diffuse comet bright but low surface brightness
L44	NGC1333	7/2/2000	03 29.2	+31 25	weird reflection nebula
*L45	NGC2188	12/17/2001	06 10.0	−34 06	Columba comet-tail galaxy
L46	NGC1851	12/17/2001	05 14.0	−40 03	Columba oval globular cluster
L47	NGC1679	12/17/2001	04 50.0	−31 59	Caelum, round galaxy
L48	NGC7582	12/17/2001	23 18.4	−42 22	Grus galaxy
L49	NGC1941	07/22/2001	04 03.4	+51 19	reflection nebula
L50	NGC1579	07/22/2001	04 30.2	+35 16	reflection nebula
L51	Arp 321	1985, 2001	09 38.9	−04 52	"Larry, Moe, and Curly" galaxies in Hydra
L52	NGC3115	date unknown	10 05.2	−07 43	spindle galaxy
L53	NGC752	07/04/2002	01 57.8	+37 41	very large open cluster in Andromeda
L54	NGC5128	12/19/2001	13 25.5	−43 01	Centaurus A
L55	NGC7664	06/09/2002	23 26.6	+25 04	galaxy in Pegasus
L56	NGC270	06/09/2002	00 50.6	−08 39	galaxy in Cetus
L57	NGC7723	06/09/2002	23 38.8	−12 58	galaxy in Aquarius
*L58	NGC772	06/09/2002	01 59.3	+19 01	round, comet-like galaxy
L59	NGC524	06/09/2002	01 24.8	+09 32	round galaxy
L60	NGC628 M74	06/09/2002	01 36.6	+15 47	diffuse galaxy
L61	NGC246	06/18/2002	00 47.0	−11 53	Cetus Ring; planetary nebula in Cetus
L62	NGC898	06/19/2002	02 23.3	+41 57	very elongated and faint galaxy in Andromeda
L63-P	NGC4656	07/06/2002	12 44.0	+32 10	"Hummingbird" galaxy found photographically

No.	Other designation	Spotted on	α (Right Ascension)	δ (Declination)	Comments[a]
			(2000.0) Coordinates		
L64	NGC404	07/04/02	01 09.4	+35 43	round galaxy near β Andromedae
L65	NGC4374 M84	07/03/02	12 25.1	+12 53	close to M86
L66	NGC4406 M86	07/03/02	12 26.2	+12 57	rich field of galaxies
L67	NGC4594 M104	07/03/02	12 40.0	−11 37	sombrero galaxy
L68	NGC1023	07/04/02	02 40.4	+39 04	very elongated galaxy in Perseus
L69-P	LW J2204.4+4509	01/02/2000	22 04.0	+45 09	"Wendee's Ring" of faint stars, found photographically
L70-P	LW J2109.0+0618	12/25/2000	21 09.0	+0618	"Equuleus S"-shaped asterism, found photographically
L71	LW J2340.6+5618	05/03/2001	23 40.6	+56 18	"Nanette's River" long chain of stars
L72	V Hydrae	11/22/1984	10 51.6	−21 15	reddest star I have found during searching
L73	TV Corvi	03/23/1990	12 20.4	−18 27	Tombaugh's star, found by searching archives
L74	NGC3623 M65	3/3/1967	11 18.9	+13 05	bright galaxy
L75	NGC3372	2/9/2000	10 45.1	−59 41	Eta Carinae Nebula
L76	NGC6618 M17	date unknown	18 20.7	−16 10	my favorite Messier object
L77	NGC2237	date unknown	06 32.3	+05 03	Rosette Nebula
*L78	NGC2261	date unknown	06 39.2	+08 44	Hubble's variable, comet-shaped
L79	NGC3628	date unknown	11 20.3	+13 36	large edge-on
*L80	NGC185	date unknown	00 39.0	+48 20	small cometary, M31 companion
*L81	NGC147	date unknown	00 33.2	+48 30	even fainter M31 companion, cometary appearance
L82	NGC5634	date unknown	14 29.6	−05 59	Virgo globular cluster
L83	NGC5638	date unknown	14 29.7	+03 14	comet-like galaxy in Virgo; mag. 11
L84	NGC6712	date unknown	18 53.1	−08 42	Scutum faint globular cluster
L85	IC1396	1966	21 39.1	+57 30	open cluster with triple and double star
L86	NGC224 M31	07/04/1966	00 42.7	+41 16	Andromeda Galaxy
L87	NGC5194 M51	07/13/1966	13 29.9	+47 12	whirlpool galaxy, a favorite
L88	NGC3587 M97	07/13/1966	11 14.8	+55 01	Owl Nebula near β Uma
L89	NGC3556 M108	date unknown	11 11.5	+55 40	galaxy near β Uma

No.	Other designation	Spotted on	(2000.0) Coordinates		Comments[a]
			α (Right Ascension)	δ (Declination)	
L90	NGC3992 M109	date unknown	11 57.6	+53 23	galaxy near γ Uma
L91	NGC598 M33	08/14/1966	01 33.9	+30 39	bright 1st Galaxy in Triangulum
L92	NGC253	10/1979	00 47.6	−25 17	Caroline Herschel's galaxy
L93	NGC2683	10/26/1979	08 52.7	+33 25	comet-like galaxy in Lynx
L94	NGC6723	11/05/1979	18 59.6	−36 38	Sagittarius globular cluster
L95	NGC1999	11/15/1979 04/09/2000	05 36.5	−06 42	diffuse nebula in Orion around a star; resembles planetary
L96	NGC2681	11/15/1979	08 53.5	+51 19	galaxy Ursa Major
L97	NGC5139	05/09/1980	13 26.8	−47 29	Omega Centauri
L98	NGC6709	11/13/1984	18 51.5	+10 21	open cluster in Aquila my first comet find in same field
L99	NGC949	06/15/1997	02 30.8	+37 08	very cometary galaxy in Triangulum
L100	NGC1491	07/19/1999	04 03.4	+51 19	near a star
L101	NGC3766	02/08/2000	11 36.1	−61 37	open cluster Centaurus
L102	IC2602	02/08/2000	10 43.2	−64 24	big nearby open cluster in Carina
L103	NGC3621	05/06/2000	11 18.3	−32 49	"Frame Galaxy" – encased in some stars
L104	NGC104	06/18/2001	00 24.1	−72 05	very bright globular
L105	NGC362	06/18/2001	01 03.2	−70 51	globular cluster in Tucana
L106	NGC4038/4039	2001	12 01.9	−19 52	Antennae or Ring Tail: colliding galaxies in Corvus
L107	NGC4361	2001	12 24.5	−18 48	planetary nebula in Corvus
L108	Σ1604	06/30/2002	12 09.5	−11 52	nice triple star in Corvus
L109	NGC936	07/09/2002	02 27.6	−01 09	Cetus elongated galaxy, bright core
L110	NGC6760	07/18/2002	19 11.2	+01 02	Aquila globular cluster
L111	NGC1068 M77	07/18/2002	02 42.7	−00 01	Cetus galaxy, very bright core
L112	NGC7023	08/13/2002	21 00.5	+68 10	most unusual appearing nebula with dust; star at edge

No.	Other designation	Spotted on	(2000.0) Coordinates		Comments[a]
			α (Right Ascension)	δ (Declination)	
L113	TU Geminorum	08/15/2002	06 10.9	+26 01	bright semi-regular red variable star mag. range 6.2–8.6
*L114	NGC1637	08/16/2002	04 41.5	−02 51	Eridanus comet-like round galaxy, mag. 10.9, diam. 3.9′
L115	NGC1788	08/16/2002	05 06.9	−03 21	Orion bright reflective nebula.
L116	NGC2158	08/17/2002	06 07.5	+24 06	compact open cluster near M35, looks nebulous at low power
L117	NGC6093 M80	08/25/2002	16 17.0	−22 59	compact globular cluster looked fuzzy when first sighted in 1980s
L118	Great Star Cloud	08/25/2002	18 03.4	−27 54	Sagittarius star cloud, stunning crowds of stars and dust lanes. Barnard 86 is at the west edge of 6520.
L119	NGC6603 M24	08/26/2002	18 16.9	−18 29	small Sagittarius star cloud, includes open cluster, fat and dense with faint stars
L120	NGC6451	08/26/2002	17 50.7	−30 13	small Scorpius open cluster,
			17 45.6	−28 56	~2 deg. SE from Galactic Center
L121	NGC5846	08/26/2002	15 06.4	+01 36	Virgo roundish galaxy
L122	NGC6826	08/29/2002	19 44.8	+50 31	blinking planetary nebula
L123	NGC1600	09/04/2002	04 31.7	−05 05	Eridanus round galaxy, diffuse
L124	NGC2174	09/04/2002	06 09.7	+20 30	large field of dust – diam. 40′. Barely noticeable but large area of brightening
L125	NGC2023	09/04/2002	05 41.6	−02 14	complex includes IC434 Barnard 33, the Horsehead
L126	NGC2359	10/01/2002	07 18.6	−13 12	Thor's helmet. Canis Minor bright nebula, involved with a gas bubble blown out from a Wolf–Rayet star

No.	Other designation	Spotted on	α (Right Ascension)	δ (Declination)	Comments[a]
			(2000.0) Coordinates		
L127	16/17 Draconis	10/05/2002	16 36.2	+52 55	similar to Epsilon Lyrae, but only one star is binary
L128	IC5020 PGC64845	10/07/2002	20 30.6	−33 29	found photographically. Galaxy with a line of foreground stars; part looks like question mark. Photographically determined mag. 13
L129	UGC 5373	10/08/2002	10 00.0	+05 20	Sextans B. Local group member. Extremely wide, about 1/4 deg.
*L130	NGC3198	10/08/2002	10 19.9	+45 33	very elongated galaxy; mag. 10.3, diam. 8′
*L131	NGC2964–2968	10/09/2002	09 42.9	+31 51	NGC 2964 brighter than 2968; mag. 11.3, diam. 2.7′
L132	NGC3432	10/09/2002	10 52.5	+36 37	Leo Minor galaxy, elongated with star at S. side and star at W. end; mag. 11.3, diam. 6.6′
*L133	NGC3070	10/11/2002	09 58.0	+10 22	Leo round galaxy, diffuse; faint companion 3069 not seen; mag. 12.3, diam. 1.3′
L134	NGC4319	10/14/2002	12 21.7	+75 19	galaxy at tail of Draco. Quasar Markarian 205 to S. easily visible with 9-mm eyepiece and averted vision. Same field as 4291 and 4386
L135	NGC4256	10/14/2002	12 18.7	+65 54	Draco edge-on galaxy with bright core; mag. 11.9, diam. 4.2′
*L136	NGC3738	10/14/2002	11 35.8	+54 31	Ursa Major near M97; mag. 11.7, diam. 2.3′. Paired with 3756; mag.11.5, diam. 4.1′
*L137	NGC3718	10/14/2002	11 32.6	+53 04	Ursa Major elongated galaxy; mag.10.8, diam. 7.9′

No.	Other designation	Spotted on	(2000.0) Coordinates		Comments[a]
			α (Right Ascension)	δ (Declination)	
*L138	NGC3953	10/14/2002	11 53.8	+52 20	see L90 M109, similar galaxy; very elongated with bright core
L139	NGC4449	10/18/2002	12 28.2	+44 06	Canes Venatici galaxy, bright core; mag. 9.6, diam. 6.1′
L140	NGC4485	10/18/2002	12 30.5	+41 42	Canes Venatici elongated galaxy; mag. 11.9, diam. 2.2′
L141	NGC4565	10/18/2002	12 36.3	+25 59	beautiful edge-on galaxy; mag. 9.6, diam. 16.2′
*L142	NGC4274	10/18/2002	12 19.8	+29 37	Coma Berenices elongated galaxy; bright core; mag. 10.4, diam. 6.6′. 4278 (mag. 10.2, diam. 3.6′) and 4314 (mag. 10.5, diam. 4.8′) are nearby but neither are as cometary
*L143	NGC4559	10/18/2002	12 36.0	+27 58	Coma Berenices elongated galaxy; mag. 10.0, diam. 10.5′
*L144	NGC4501 M88	10/18/2002	12 32.0	+14 25	very elongated galaxy, bright core; mag. 9.5, diam. 5.8′
*L145	NGC4473	10/18/2002	12 29.8	+13 26	Coma Berenices very elongated galaxy with bright core; mag. 10.2, diam. 4.2′
*L146	NGC4472 M49	10/18/2002	12 29.8	+08 00	Virgo round galaxy; bright core; mag. 8.4, diam. 8.9′
L147	NGC6638	10/24/2002	18 30.9	−25 30	Sagittarius globular, added by Dean Koenig
L148	R Leporis	10/31/2002	04 59.6	−14 48	redder when fainter
*L149	NGC6553	11/04/2002	18 09.3	−25 54	Sagittarius globular, faint and large; mag. 8.3, diam. 7.9′
L150	NGC6523 M8	11/04/2002	18 03.8	−24 23	Comet Levy C/1990 K1 passed over M8 in 1990; mag. 5.0, diam. 76.0′

No.	Other designation	Spotted on	α (Right Ascension)	δ (Declination)	Comments[a]
			(2000.0) Coordinates		
L151	NGC6611 M16	11/04/2002	18 18.8	−13 47	Eagle Nebula with cluster; mag. 6.0, diam. 69.0′
L152	NGC6514 M20	11/04/2002	18 02.3	−23 02	Trifid Nebula, colors in 16-inch telescope; mag. 6.3, diam. 24′
L153	NGC7293	9/11/1982	22 29.6	−20 48	Helix Nebula, looks ghostlike; mag. 7.3, diam. 860′
L154	NGC7009	11/04/2002	21 04.2	−11 22	Saturn nebula, blue!; mag. 8.0, diam. 3.0′
L155	NGC6981 M72	11/04/2002	20 53.5	−12 32	Aquarius globular near 7009; mag. 9.4, diam. 5.8′
L156	NGC6934	11/04/2002	20 34.2	+07 24	Delphinus globular; mag. 8.9, diam. 5.8′
L157	LW J1948.3+3744	11/04/2002	19 48.3	+37 44	The Cane. Asterism. 3 stars in handle; 4 in curve; 12 in straight line; triangle at bottom. Position is of brighest star HIP 97435, and most of the line extends 1/2 deg. N. and E. of it. The bright star is 1 deg. S. and a little W. of 19 Cygni. In wide field, Cane is framed mostly on W. side by a "semicircle" of bright stars including 19 Cygni.
*L158	"Castor Cluster"	11/04/2002			
	IC2196		07 34.1	+31 24	Gemini galaxy; mag. 14.0, diam. 1.4′;
	IC2197		07 34.3	+31 24	Gemini galaxy; mag. 14.0; diam. 0.4′;
	IC2194		07 33.7	+31 19	Gemini galaxy; mag. 15.0, diam. 0.9′; these were originally spotted during the 1980s
L159	NGC2264	11/04/2002	06 41.1	+09 53	Xmas tree cluster with nebulosity; mag. 3.9, diam. 1 deg.
L160	NGC2254	11/04/2002	06 36.0	+07 40	"Mountains in the Sky" Monoceros open cluster and star chain

No.	Other designation	Spotted on	(2000.0) Coordinates		Comments[a]
			α (Right Ascension)	δ (Declination)	
*L161	NGC2245	11/04/2002	06 32.7	+10 10	comet-like bright nebula, looks like Hubble's Variable Nebula
L162	NGC2252	11/04/2002	06 35.0	+05 23	Monoceros open cluster, looks like a rope of stars
*L163	NGC2775	11/04/2002	09 10.3	+07 02	Cancer round galaxy; mag. 10.1, diam. 4.2′
*L164	NGC3486	11/04/2002	11 00.4	+28 58	Leo Minor round galaxy; mag. 10.5, diam. 6.9′
*L165	NGC3245	11/04/2002	10 27.3	+28 30	Leo Minor elongated galaxy; mag. 10.8, diam. 3.1′
*L166	NGC3344	11/04/2002	10 43.5	+24 55	Leo Minor round galaxy, bright core, star nearby; mag. 9.9, diam. 6.9′. Don Machholz quotes Peltier as saying this looks like a comet. George Alcock echoes that its cometary appearance is enhanced by the nearby star
L167	NGC3310	11/04/2002	10 38.7	+53 30	Ursa Major round galaxy; mag. 10.8, diam. 4.2′
L168	NGC3242	11/04/2002	10 24.8	−18 38	"Ghost of Jupiter" planetary nebula; mag. 7.7, diam. 4.0′
*L169	NGC2986	11/04/2002	09 44.3	−21 17	Hydra round galaxy, bright core; mag. 10.8, diam. 3.1′
L170	NGC4651	11/04/2002	12 43.7	+16 24	Coma Berenices round galaxy, bright core; mag. 10.8, diam. 3.9′
L171	NGC4450	11/04/2002	12 28.5	+17 05	Coma Berenices elongated galaxy, bright core; mag. 10.1, diam. 5.0′
*L172	NGC4689	11/04/2002	12 47.8	+13 46	Coma Berenices elongated galaxy; mag. 10.9, diam. 4.2′

No.	Other designation	Spotted on	α (Right Ascension)	δ (Declination)	Comments[a]
			(2000.0) Coordinates		
*L173	NGC4548 M91	11/04/2002	12 35.4	+14 30	Coma Berenices elongated galaxy, bright core; mag. 10.2, diam. 4.2′
*L174	NGC4649 M60	11/04/2002	12 43.7	+11 33	Virgo galaxy; fainter one nearby; mag. 8.8, diam. 7.1′
L175	NGC4486 M87	11/04/2002	12 30.8	+12 24	Virgo elliptical galaxy; mag. 8.6, diam. 6.9′
*L176	NGC4579 M58	11/04/2002	12 37.7	+11 49	Virgo galaxy; mag. 9.8, diam. 4.7′
L177	NGC4552 M89	11/04/2002	12 35.7	+12 33	Virgo galaxy; mag. 9.8, diam. 5.0′
*L178	NGC4596	11/04/2002	12 39.9	+10 11	Virgo elongated galaxy, bright core; mag. 10.4, diam. 3.9′
*L179	NGC4535	11/04/2002	12 34.3	+08 12	Virgo round galaxy, bright core; mag. 10.0, diam. 6.9′
*L180	NGC4303 M61	11/04/2002	12 21.9	+04 28	Virgo spiral galaxy; low surface brightness; first spotted 1967; mag. 9.7, diam. 6.3′
*L181	NGC3887	11/04/2002	11 47.1	−16 51	Crater round galaxy; mag. 10.6, diam. 3.1′
L182	NGC4636	11/10/2002	12 42.8	+02 41	Virgo galaxy, bright core; mag. 9.5, diam. 5.8′
*L183	NGC4818	11/10/2002	12 56.8	−08 31	Virgo very elongated galaxy; barely visible in 16-inch telescope; mag. 11.1, diam. 4.2′
*L184	NGC5147	11/10/2002	13 26.3	+02 06	Virgo round galaxy, faint; mag. 11.8, diam. 1.8′
L185	NGC5248	11/10/2002	13 37.5	+08 53	Boötes round galaxy; mag. 10.3, diam. 6.1′
*L186	NGC5371	11/11/2002	13 55.7	+40 28	Canes Venatici round galaxy; mag. 10.6, diam. 4.2′
*L187	NGC5020	11/11/2002	13 12.6	+12 36	Virgo elongated galaxy, bright core; mag. 11.7, diam. 3.1′

| No. | Other designation | Spotted on | (2000.0) Coordinates | | Comments[a] |
			α (Right Ascension)	δ (Declination)	
*L188	NGC4591	11/11/2002	12 39.3	+06 01	Virgo elongated galaxy, faint; 16-mm eyepiece; mag. 13.0, diam. 1.5′
L189	NGC5127	11/12/2002	13 23.8	+31 34	Canes Venatici round galaxy, bright core, faint; mag. 11.9, diam. 2.7′
L190	NGC4956	11/12/2002	13 05.1	+35 11	Canes Venatici round galaxy; 16-mm eyepiece; mag. 12.4, diam. 1.4′
L191	NGC4772	11/12/2002	12 53.5	+02 10	Virgo galaxy; mag. 11.0, diam. 3.3′
*L192	NGC4536	11/12/2002	12 34.5	+02 11	Virgo very elongated galaxy; mag. 10.6, diam. 7.4′
*L193	NGC4129	11/12/2002	12 08.9	−09 02	Virgo elongated galaxy; 16-mm eyepiece; mag. 12.5, diam. 2.2′
L194	NGC281	11/13/2002	00 52.8	+56 37	Cassiopeia open cluster with well-shaped nebulosity; found photographically via Schmidt camera; mag. ~7, diam. 35′
L195	NGC2419	11/13/2002	07 38.1	+38 53	Shapley's Intergalactic Wanderer; Lynx globular cluster; mag. 0.3, diam. 4.1′
L196	NGC5694	*circa* 1990	14 39.6	−26 32	Tombaugh's Cluster, looks like a snowball at the end of a shovel of nearby stars
L197	NGC5907	10/24/1982	15 15.9	+56 19	Draco very long galaxy; mag. 10.4, diam. 12.3′
L198	NGC2437 M46	03/15/1983	07 41.8	−14 49	Puppis open cluster; mag. 6.1, diam. 27′
	NGC2438		07 41.8	−14 44	planetary nebula in foreground of M46; mag. 10.8, diam. 1.1′
L199	NGC4567/4568	03/15/1983	12 36.5	+11 15	Virgo spectacular Siamese Twins galaxies; 4567 mag. 11.3, diam. 3.0′; 4568 mag. 10.8, diam. 4.6′

No.	Other designation	Spotted on	α (Right Ascension)	δ (Declination)	Comments[a]
			(2000.0) Coordinates		
L200	Fornax cluster	1983			Cluster of several galaxies in same field
	NGC1380		03 36.5	−34 59	Fornax galaxy; mag. 11, diam. 4.9′
	NGC1399		03 38.5	−35 27	mag. 9.9, diam. 3.2′
	NGC1404		03 38.9	−35 35	mag. 10.3, diam. 2.5′
*L201	NGC3865	03/05/1989	11 44.9	−09 14	Crater galaxy, faint and diffuse; mag. 13, diam. 2.3′
*L202	NGC5427	03/05/1989	14 03.4	−06 02	Virgo galaxy; mag. 11.4, diam. 2.5′
*L203	NGC5668	03/05/1989	14 33.4	+04 27	Virgo galaxy; mag. 11.5, diam. 3.3′
*L204	NGC5850	03/05/1989	15 07.1	+01 33	Virgo galaxy; mag. 11.0, diam. 4.3′
*L205	NGC6106	03/05/1989	16 18.8	+07 25	Hercules galaxy; mag. 12.2, diam. 2.6′
*L206	NGC6118	03/05/1989	16 21.8	−02 17	Serpens galaxy; mag. 12.0, diam. 4.7′
*L207	NGC6384	03/05/1989	17 32.4	+07 04	Ophiuchus galaxy; mag. 10.6, diam. 6.0′
*L208	NGC6426	03/05/1989	17 44.9	+03 00	Ophiuchus globular cluster; mag. 11.2, diam. 3.2′
*L209	NGC3049	03/06/1989	09 54.8	+09 16	Leo galaxy
*L210	NGC4685	11/13/2002	12 47.1	+19 28	"Winking Galaxy", Coma Berenices round galaxy with very bright core; concentrate on the core and the rest of the galaxy disappears!; 16-mm eyepiece; mag. 12.6, diam. 1.5′
*L211	NGC4779	11/13/2002	12 53.8	+09 44	Virgo round galaxy; 16-mm eyepiece; mag. 12.4, diam. 2.0′
*L212	NGC4795	11/13/2002	12 55.0	+08 04	Virgo round galaxy; 16-mm eyepiece; mag. 12.1, diam. 1.8′
*L213	NGC4623	11/13/2002	12 42.2	+07 41	Virgo very elongated galaxy; 16-mm eyepiece; mag. 12.2, diam. 2.1′

No.	Other designation	Spotted on	(2000.0) Coordinates		Comments[a]
			α (Right Ascension)	δ (Declination)	
*L214	NGC4713	11/13/2002	12 50.0	+05 19	Virgo elongated galaxy; 16-mm eyepiece; mag. 11.7, diam. 2.6′
*L215	NGC4688	11/13/2002	12 47.8	+04 20	Virgo very large, low surface brightness galaxy; looks like a diffuse comet; 16-mm eyepiece; mag. 11.9, diam. 3.1′
L216	NGC4590 M68	11/15/2002	12 39.5	−26 45	Hydra globular cluster; diffuse at low altitude; mag. 8.2, diam. 12.0′
L217	NGC3923	11/15/2002	11 51.0	−28 48	Hydra elongated galaxy, bright core; mag. 9.8, diam. 5.8′
L218	NGC3201	11/15/2002	10 17.6	−46 25	Vela globular cluster; large; mag. 6.8, diam. 18.2′
L219	Hydra 1 cluster	11/17/2002			includes:
	NGC3309		10 36.6	−27 31	brightest in rich cluster of galaxies,
	NGC 3311		10 36.7	−27 32	
	NGC3312		10 37.0	−27 34	
	NGC3314		10 37.4	−27 41	actually 2 spirals, one directly in front of the other as seen through HST
	NGC3316		10 37.6	−27 36	
L220	NGC6910	12/17/2002	20 23.1	+40 47	Cygnus open cluster near γ Cygni; mag. 7.4, diam. 8.0′
		12/25/2002			encountered visually

[a] deg. = degree; diam. = diameter; mag. = magnitude.

Index

Bold arabic numbers refer to figures; roman numbers refer to color plates